THE OPTICS BOOK

Fun Experiments with Light, Vision & Color

Shar Levine &
Leslie Johnstone

Illustrations by Jason Coons

Sterling Publishing Co., Inc.
New York

To Peter and Michael, for your friendship and, as always, to my family.—S. L.
For Chris, Nick, and Megan, who make it all worthwhile.—L. J.
To Glossé J. and Finch.—J. C.

Acknowlegments

Our thanks to Jeff Connery of Printed Light Photography, Burnaby, B. C., Canada, for the great work and the sushi on cold Vancouver winter days and to Sean Lavery and the Science Department of Point Grey Secondary School, Vancouver, B. C. Also to Bryn Hughes for help with the pinhole camera.

Designed by Judy Morgan.

Photos on pages 1,4, 8 (lower left), 19, 24 (lower left and right), 25, 28, 32, 35, 37, 59, 62 right, and 63 courtesy of Central Supply Company, Franklin Park, Illinois. Photos on page 60 courtesy of Edmund Scientific Co., Barrington, New Jersey. Other photos by Jeff Connery of Printed Light Photography and the authors.

Library of Congress Cataloging-in-Publication Data

Levine, Shar, 1953-
The optics book: fun experiments with light, vision & color / Shar Levine & Leslie Johnstone; illustrated by Jason Coons.
p. cm.
Includes index.
Summary: Explores the properties of light and color by means of experiments and analysis of various optical instruments including periscopes and telescopes.
ISBN 0-8069-9947-0
1. Optics—Juvenile literature. 2. Optics—Experiments—Juvenile literature. 3. Color—Experiments—Juvenile literature. 4. Optical instruments—Juvenile literature. [1. Optics—Experiments. 2. Color—Experiments. 3. Experiments. 4. Optical instruments.]
I. Johnstone, Leslie. II. Coons, Jason. ill. III. Title.
QC360.L48 1998
535—dec21
98-26732
CIP
AC

3 5 7 9 10 8 6 4 2

First paperback edition published in 1999 by
Sterling Publishing Company, Inc.
387 Park Avenue South, New York, N.Y. 10016

Distributed in Canada by Sterling Publishing
% Canadian Manda Group, One Atlantic Avenue, Suite 105
Toronto, Ontario, Canada M6K 3E7
Distributed in Great Britain and Europe by Chris Lloyd
463 Ashley Road, Parkstone, Poole, Dorset, BH14 0AX, England
Distributed in Australia by Capricorn Link (Australia) Pty Ltd.
P.O. Box 6651, Baulkham Hills, Business Centre, NSW 2153, Australia
Printed in Hong Kong

Sterling ISBN 0-8069-9947-0 Trade
0-8069-9942-X Paper

CONTENTS

PREFACE

Optics is the study of visible light and vision and the instruments that use light. What light is, how it acts, and how we can see light are all important parts of the science of optics. We often take light for granted because it is all around us. Try this: close the blinds in the room you are in and turn off all the lights. Now shut the door so that the room is completely dark. How does it feel to be surrounded by blackness? That's what our world would be like without the light from the Sun.

From the earliest beginnings, people have been fascinated with light. Early cave drawings involved pictures of the Sun and the Moon. People even prayed to the Sun and worshiped gods named after the Sun. These people weren't too sure that the Sun was going to rise the next day. Their rituals tried to ensure that it would.

For thousands of years people have tried to explain light. The ancient Greeks were among the first to try to describe light. Around 500 B.C., Pythagoras, who is best known for his mathematical theorem, believed that light came out of the eye, allowing the viewer to see objects. Two hundred years later, Aristotle, another famous Greek philosopher, described light as moving outward from the source like waves in water. In the 11th century, the Arabic scientist Alhazen stated that light rays go from the objects into the eyes of the observer. In the 17th century, the Dutch scientist Christiaan Huygens

described some experiments to show that light acts like a wave. Around the same time, Isaac Newton developed a theory of light which treated light as a stream of particles.

This clash of ideas, in which light is thought of either as some kind of particle or ray or as some type of wave, continued for over 2000 years. Today, scientists know that light is a form of energy. Light is part of a wide range of electromagnetic waves that includes radio waves and x-rays. It has some properties of both waves and of particles.

Scientists use models to explain the ways they think things work. A model is an object or idea used to understand something else. A model of a house might be used to help you learn the location of the rooms, for example. In this book, you will read about some models of light. These are ways to understand light. If one model can't explain an observation, scientists come up with a new model that can. Sometimes we will use the ray model to understand light better and sometimes we will use the wave model of light.

NOTE TO PARENTS AND TEACHERS

This book contains experiments that involve color. Some of the readers of the book may not be able to see some of the differences in color that are described. They may wonder what the book is trying to explain. It is possible that some readers may have a visual condition called color blindness. This usually involves difficulty in distinguishing between tints or shades of red and green. These colors may be confused or may look like gray to the affected person. Color blindness is an inherited condition that occurs because some color receptors are missing in the retina. Red-green color blindness is the most common kind. It is more common in males than in females.

If you notice that someone is having difficulty seeing some colors, plan your project so that activities with the person don't require distinguishing between those colors. Speak with the person privately, and offer reassurance that there is nothing to be ashamed of. If the person wants to learn more about color blindness, suggest checking with a physician.

SAFETY FIRST

Before you do any of the activities in this book, it is important to consider your safety. Below are a few important rules to follow.

Do's

1. Ask an adult before handling any materials or equipment.

2. Have an adult handle all sharp objects such as knives or scissors.

3. Tie back long hair and avoid wearing clothing with long loose sleeves while you work; they could knock things over.

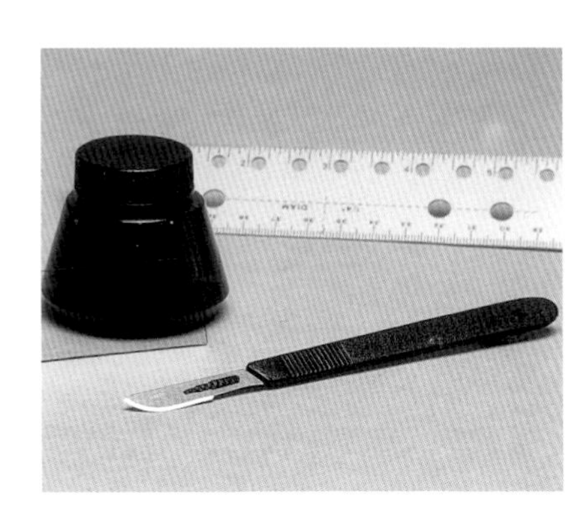

4. Wash your hands after handling any chemicals.

5. Read the entire activity before beginning, and put the supplies needed nearby.

6. Tell an adult immediately if you hurt yourself in any way.

7. Keep all supplies, materials and chemicals out of the reach of very young children.

8. Keep your work area neat and clean and immediately clean up any spills.

9. Safely dispose of all chemicals and materials you can't reuse when you are finished with them.

Don'ts

1. Do not taste, eat, or drink any of the experiments.

2. Never look at the Sun or other strong light sources through your lenses.

3. Never leave magnifying lenses in direct sunlight.

LIST OF MATERIALS

aluminum foil

scissors

ruler

2 clear plastic protractors

several pieces of cardboard (see projects for more details)

cellophane tape

masking tape

wooden spoon

modeling clay

toothpicks

large shoebox and lid

craft knife

drawing compass

flashlight

2 mirrors, size 4 × 5 inches (10 × 12 cm)

magnifying mirror

shiny metal spoons

water

salt

glass containers and jars

polarizing filters

diffraction grating

wax paper

prism

microscope slides

clear plastic fork

tempera paints

elastic bands

colored markers

colored filters or colored cellophane: red, green, blue, yellow, magenta, cyan (blue-green)

clear cellophane

two convex lenses of different thicknesses, about 1½ inches (4 cm) in diameter

concave lens

quart- or litre- sized milk container

piece of silver Mylar™

watch with a second hand

Light Rays & Reflection

Light travels outwards from a light source and spreads out over a larger and larger area as it travels. If nothing is placed in the way, or acts to change it, light will travel in straight lines. You will have seen this if you've ever pointed a flashlight beam at something on a dark night. The flashlight beam goes straight towards wherever it is shone. We can think of light as traveling outward in narrow beams or rays. This idea, called the ray model of light, can be used to explain many observations about light we will make in this section of the book.

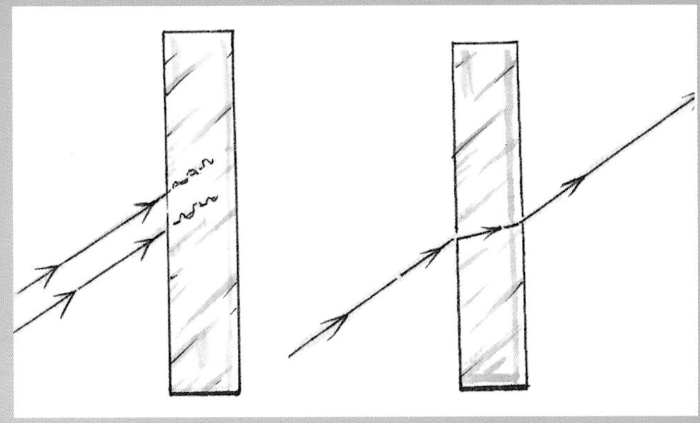

Left, absorbed; right, refracted

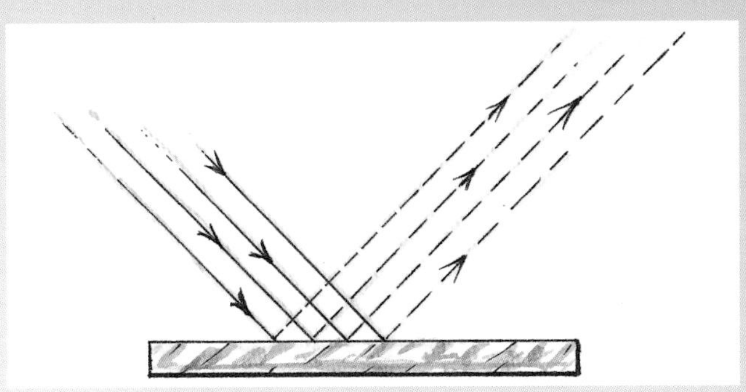

Specular reflection

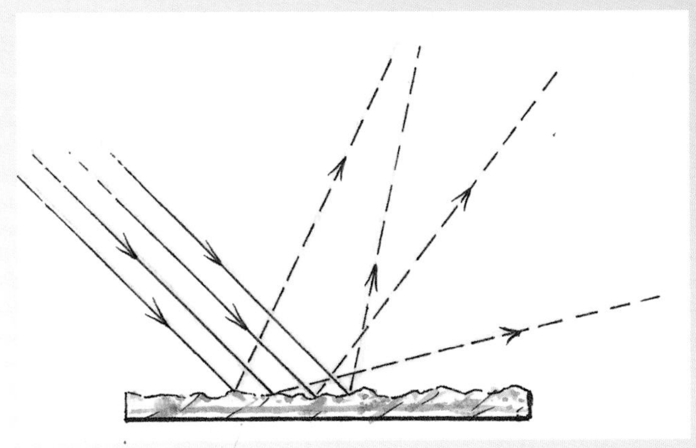

Diffuse reflection

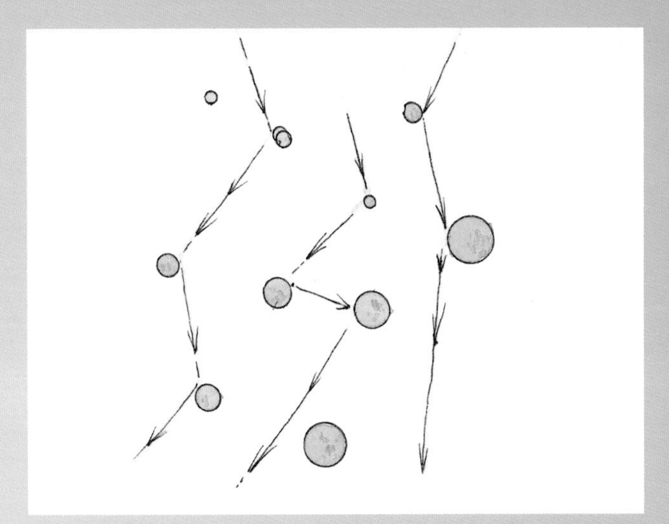

Scattered

When light hits an object, there are several possible things that can happen. Light can be reflected or bounced off the object. Light can be absorbed and cause the object to heat up. If the object is clear, or transparent, like water or window glass, light can be transmitted or can travel through the object. Light can be scattered, absorbed by tiny particles such as dust or water droplets, and quickly sent forth in another direction. When light hits most objects, some is transmitted, some is absorbed, and some is reflected.

Have you ever wondered why you can see your reflection in a mirror, or on a piece of aluminum foil or glass, but not in a piece of paper or cloth? How an object reflects light depends on the roughness of the surface of the object. A rough surface like paper or cloth reflects light in many directions at once. This is called diffuse reflection. A smooth surface like a mirror reflects light so that a thin beam of light hitting the mirror produces a thin beam of light bouncing off the mirror. This is called specular reflection.

Mirrors reflect light in interesting ways. You probably perform a simple science experiment every morning when you brush your teeth or comb your hair in front of a mirror. There you are, reflected in the mirror. Try this tomorrow morning: Hold your toothbrush in your right hand. Look in the mirror. Is the reflected person holding the toothbrush in the right or left hand? If you were looking at a real person doing what your image is doing, the toothbrush would be in the person's left hand. Confused?

Laser light reflected in a mirror.

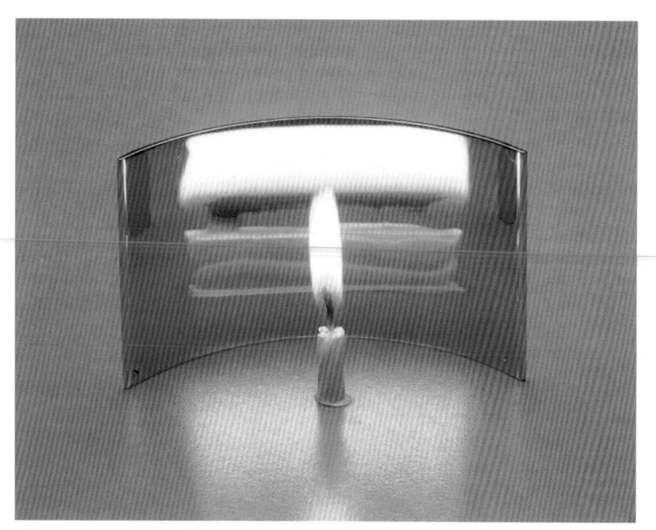

Candle reflected in concave mirror.

Plane (flat) mirrors don't invert images. This means your image in a flat mirror is right-side up, as opposed to upside down. Despite the fact that whatever you see in the mirror looks like you, it is not an accurate image of you. In fact, some mirrors can alter your image a lot. Curved mirrors can make objects appear to be bigger or smaller than they actually are. Funhouse mirrors, which make you look very tall, or short, or thin, or fat, are simply curved mirrors. Mirrors can even be used to focus or concentrate light.

In this section of the book, we'll look at some of the ways that light rays act when they are reflected.

TOASTED MARSHMALLOWS*

If you stay out too long in the Sun without sunscreen, you can cook your skin. This is also known as getting a sunburn; it isn't very good for your body. Even on cloudy days, some of the Sun's rays can be harmful. These rays are easily blocked by sunscreen or clothing. You can harness some of the Sun's strong rays and use them for something handy, like a portable oven. Here is one science experiment where you can eat the results.

**Warning: Wear a hat, sunscreen, sunglasses, and a long-sleeved shirt if you are going to be out in the Sun for any period of time. Never look directly at the Sun.*

You Will Need

aluminum foil

scissors

ruling compass

ruler

thin cardboard

tape

large plastic bowl

wooden spoon

blob of modeling clay

toothpick

marshmallow

a sunny day

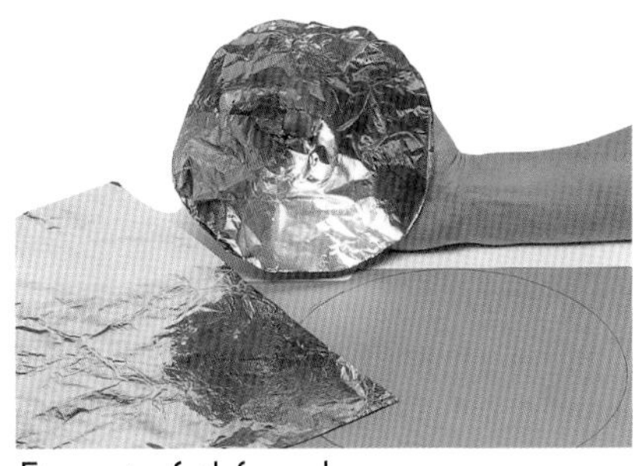

Finger in foil funnel.

What to Do

1. Cut out a circle of aluminum foil 8 inches (20 cm) across. Cut a slit into the center of the circle, starting from the edge. Make a hole in the center of the circle that is large enough to stick your index finger through. Cut a piece of thin cardboard the same size as the foil. Tape the foil, shiny side out, to the cardboard.

2. Make a funnel shape of the foil and cardboard with the foil on the inside. Tape it to keep its shape. Place your finger in the hole so it points towards the wider part of the funnel. Point your finger towards the Sun. When your finger feels warmer, remove the foil funnel. Set it aside.

3. Take a large plastic bowl and line the inside with aluminum foil, shiny side up. Smooth out the foil with the back of a wooden spoon.

4. Place a blob of modeling clay in the bottom of the bowl. Stick a toothpick upright into the modeling clay. Attach a marshmallow to the toothpick.

5. Place the bowl in a sunny place and watch what happens.

Marshmallow in foil-lined bowl.

What Happened

Your finger heated up when you pointed the foil funnel at the Sun. The marshmallow got even hotter when it was placed in the sunlight. The light from the Sun is a form of energy called electromagnetic radiation. Parts of sunlight we see as white light or colors. Parts of the Sun's radiation warm us and can cook food if they are focused.

Your foil-lined bowl concentrated the Sun's energy by reflecting the light onto a small area. This energy can be used to heat houses and swimming pools and to power solar cells and solar batteries.

The Radiometer: Light Is Energy

Light is a form of energy. Energy can be described as the ability to do work. One way of doing work is making something move. One device that shows that light can make something move is a radiometer. It uses four vanes of aluminum foil hanging at the center on a point. Each of the vanes is blackened on one side and shiny on the other side. The vanes are hung in a transparent glass bubble, from which most of the air is removed. When light shines on the radiometer, the vanes absorb more light on their dark sides than on their light sides, and the dark sides get warmer. More of the heated molecules of the gases which make up the air bounce off the vanes on the dark sides than on the shiny sides. This causes the vanes to turn. The light energy is changed into motion.

You can purchase a radiometer at at some science stores and museums. Shine a light on the glass bulb or leave it in a sunny spot, and watch as the vanes inside the bulb spin in a circle. Note: a radiometer is made of glass and can easily break if banged or dropped, so treat it gently.

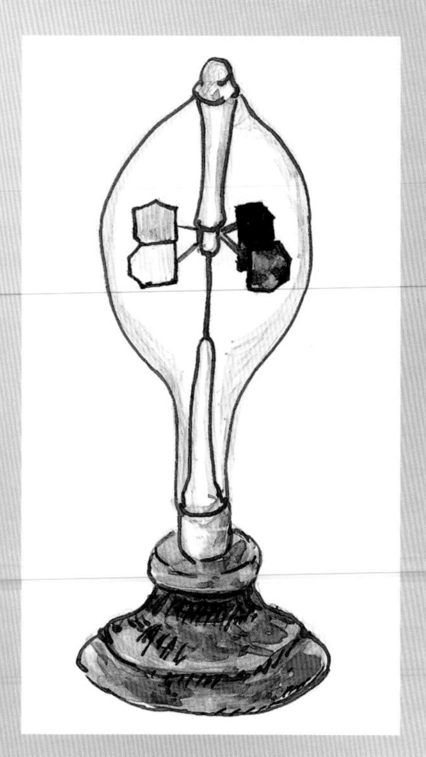

LIGHT IN A BOX

Light travels outwards in all directions from a point light source. When scientists want to study the behavior of light, they often use a light source that produces a thin beam of light traveling in only one direction. This device is called a ray box. A ray box can be used to study the behavior of light that is reflected by a mirror or refracted (bent) by a lens. Here is an easy way to make your own ray box.

You Will Need

large shoebox with lid

craft knife or scissors

cardboard about 4 × 5 inches (10 × 12 cm)

masking tape

small flashlight

What to Do

1. Use a craft knife or scissors to cut a hole 3 inches × 2 inches (7.5 × 5 cm) on one short side of a shoebox at the bottom (see diagram).

2. Cut a rectangle of lightweight cardboard 1 inch (2.5 cm) larger than the hole. Make three narrow slits, about 1/2 inch (1 cm) apart, in the cardboard (see diagram). The slits should be very narrow, just wide enough to let some light through (1/16 inch or .2 mm) and about 1 inch (2.5 cm) long. This slitted rectangle will be a baffle (a device for regulating or deflecting the intensity of light).

3. Tape the cardboard baffle over the hole in the shoebox with the slits touching the bottom edge of the box. Be sure no light is coming through any cuts except the baffle slits.

4. Place the flashlight inside the box, facing so that the light can shine out through the slits. Adjust the position of the flashlight in the box so the light rays coming out through the slits are parallel to each other.

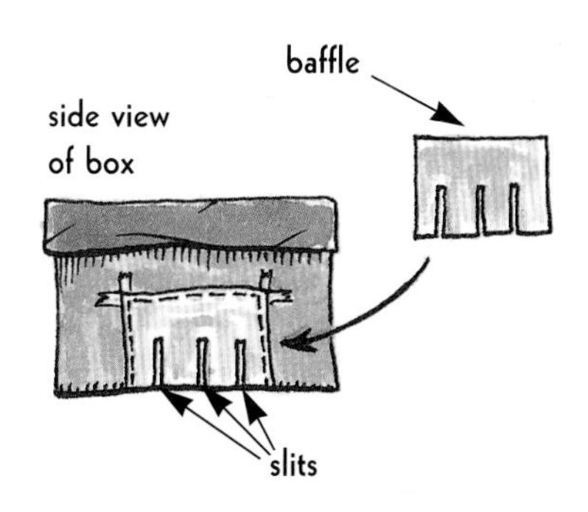

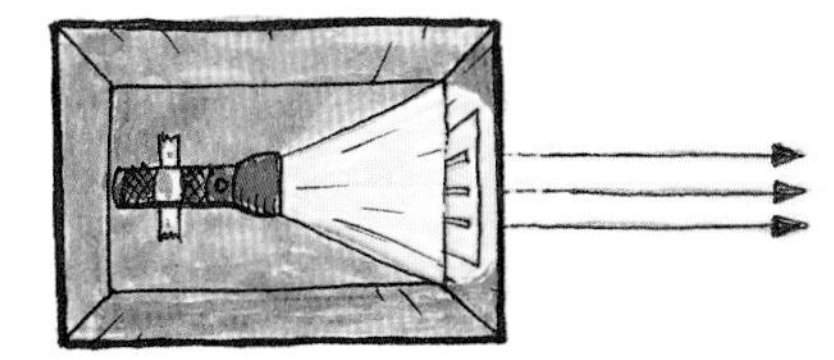

Making a ray box

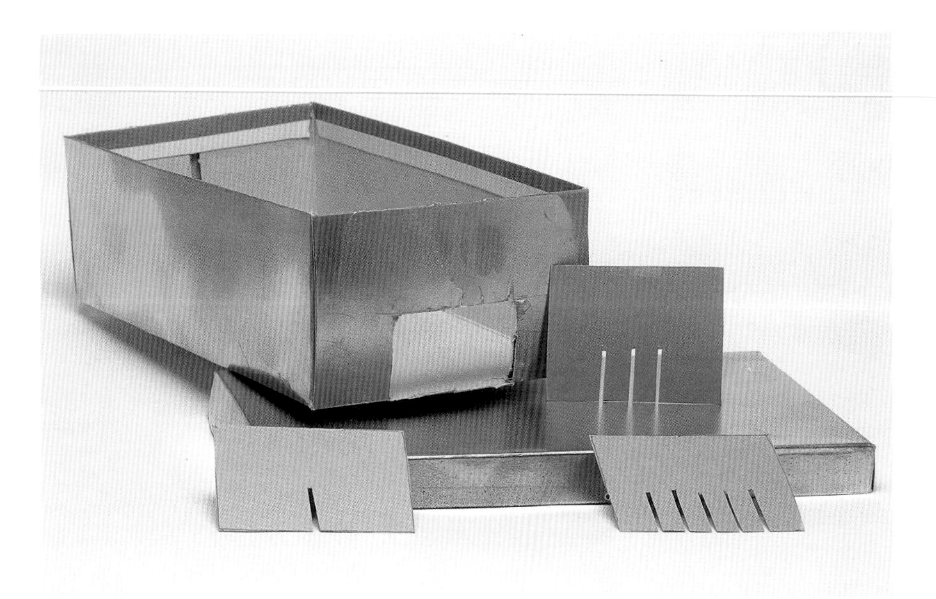

Ray box and 3 baffles.

What Happened

You have made a ray box. When you move the flashlight towards the cardboard baffle, the light rays spread outwards from the box. As the flashlight was moved away from the baffle, the light rays came closer together. If the box was big enough and the flashlight was small enough, the light rays should have appeared to be close to parallel. We will use the ray box for some of our other experiments with light.

You may need to cut a hole in the wall of the box opposite the baffle and move the flashlight handle back through the hole to get it far enough away to give parallel rays. Tape the flashlight in place in the box and close the box lid.

5. Turn off the room light and turn on the flashlight. Observe the light coming through the slits.

6. Try making another baffle, with five slits instead of three, and a baffle with only one slit. Insert each in your ray box after removing the first baffle. Do you need to move the flashlight to keep the light rays parallel?

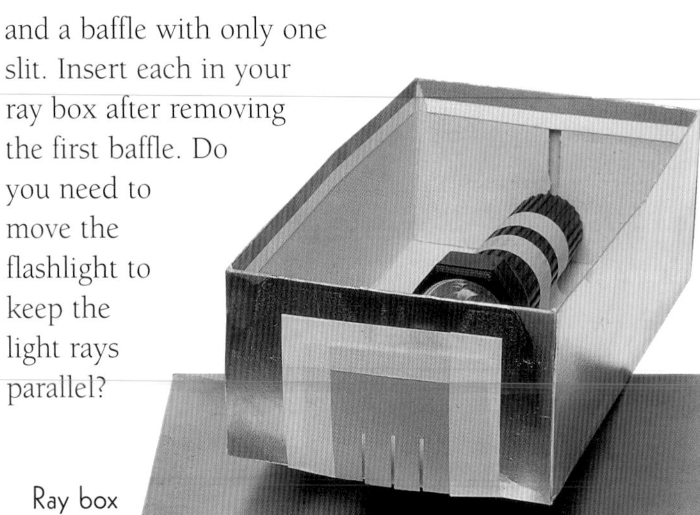

Ray box with baffle in place.

AGAINST THE LAW

There are laws of gravity. There are laws of motion. There are even laws of reflection. These are not called laws because you will be charged by the police if you break them; they are called laws because they accurately predict the way something in nature will happen. In this experiment, you will see the relationship between the angle at which light strikes a surface and its reflection.

You Will Need

ray box from Light in a Box

pencil, blue marker, and another color of marker

ruler

protractor

rectangular mirror

rectangular piece of white cardboard or paper the size of the mirror or bigger

masking tape

wall or book

notebook or paper

What to Do

1. Use a ruler to measure the cardboard; then draw a dotted pencil line down the center of the piece of cardboard or paper, perpendicular to the long edge (see photo).

2. Prop the mirror up against a wall or a book so that it stands upright and forms a right angle to the table or floor. Tape it in place if necessary.

3. Place the cardboard under the mirror so the line down the center of the cardboard is approximately in the middle of and at a right angle to the mirror (see photo). This center line represents the normal, a line at right angles to the mirror.

4. Place one end of the ruler at the end of the center line next to the mirror and, using a blue marker, draw a solid line from the end of the center line to the corner of the cardboard.

5. Find the reflected solid blue line in the mirror, and use your ruler and blue marker to extend a straight, dashed line from the edge of the reflected blue line in the mirror onto the cardboard.

6. Use a protractor to measure the angle (angle A) from the dotted pencil line to the solid blue line and record this in your notebook. This is the angle from the normal to the line that hits the mirror; it is called the angle of incidence. Now measure the angle from the normal to the dotted blue line (angle B) and record this in your notebook. This angle is the same size as the reflected angle in the mirror—it is the angle of reflection.

7. Try this again, changing the angle. Use a different colored marker for the lines you draw.

8. Using a ray box, shine a beam of light along the solid blue line. Observe how the light rays are reflected.

What Happened

When you shone the light onto the mirror along the solid blue line, you saw that it was reflect-

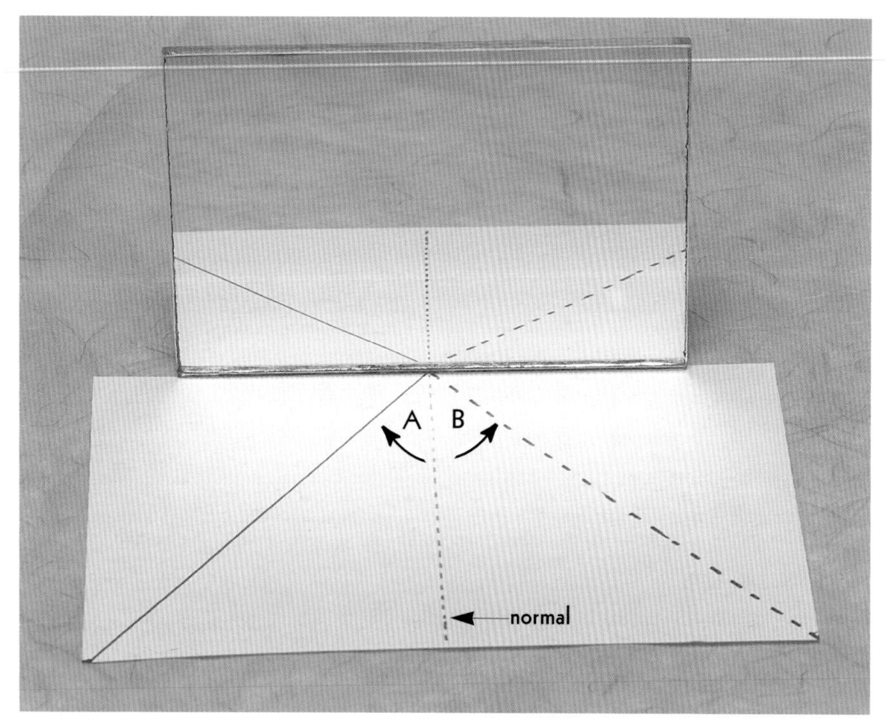

A, the angle of incidence is equal to B, the angle of reflection.

ed along the dotted blue line. Compared to the perpendicular pencil line (the normal), the angle at which the light was reflected was the same as the angle that the incident light made with the normal. This proves a law that the angle of incidence is equal to the angle of reflection.

Using the ray box allowed you to measure these angles directly. The reflection of the line in the mirror appears to be an extension of the line a reflected beam would make on the cardboard.

THE MIRROR HAS TWO FACES

In the changing room of many clothing stores there are adjustable mirrors on hinges. These mirrors help you see what your clothes will look like from behind or from the side. Many people claim these are trick mirrors because there is no way they are that big from behind! Here is an experiment on angles, mirrors, and images. And as everyone knows, mirrors never lie, or do they?

You Will Need

2 rectangular mirrors, size about 4 × 5 inches (10 cm × 12 cm)

masking tape

clear plastic protractor

coin, candle, or other small object

bit of modeling clay

paper and pencil

What to Do

1. Use masking tape to form a hinge to join the two mirrors on a long side. Stand them on their short sides (see photo).

2. Place the small object between the two mirrors, an equal distance out from each mirror. Use the blob of modeling clay to stick the object to the table so it is upright, if necessary.

3. Count the number of times the object is reflected in the mirror. Measure the angle between the mirrors by putting your protractor on top of the mirrors. Record the angle and the number of images you see.

4. Without moving the object, change the angle between the two mirrors. Count the objects reflected in the mirror again and measure the angle. Do this several times, each time making the angle larger or smaller than the one in Step 3. Count the images each time you change the angle, and record your findings.

What Happened

When you changed the angle, the number of objects reflected also changed. If the mirrors form a 90 degree angle, or right angle to each other, the light from the object is reflected twice and you should see three images. If the angle is less than 90 degrees you will see more than three images. As you decrease the angle, you see more and more images.

IMAGE-IN THAT

In The Mirror Has Two Faces, you discovered the effect that the angle between two mirrors had on the number of reflections of objects in the mirrors. What do you think would happen if the mirrors were facing each other? Would the images seem to go on forever? Here's one way to find the answer to that.

You Will Need

2 rectangular mirrors, size about 4 × 5 inches (10 cm × 12 cm)

masking tape

ruler

2 large, heavy books or two tissue boxes

coin, candle or other small object

bit of modeling clay

What to Do

1. Place each book flat on the table. Tape each mirror, shiny side out, to the side of a book. This will keep the mirrors from falling over and breaking during the experiment. Use boxes if you don't have books.

2. Arrange the books so the mirrors are parallel to and facing each other. Leave a space of about 5 inches (12 cm) between the mirrors.

3. Place an object such as a candle between the two mirrors. Use the modeling clay to hold it upright if necessary. Look at the reflections in the mirror and count how many times the object appears.

What Happened

The object was reflected by the mirrors and formed a row of images, getting smaller and smaller in the distance. When the mirrors are placed directly facing each other, the images act as if they were the original object. Not only is the object reflected, but also the images are reflected. Since a little bit of the light energy is absorbed by the mirror with each reflection, some of the smaller images seem to be fainter also.

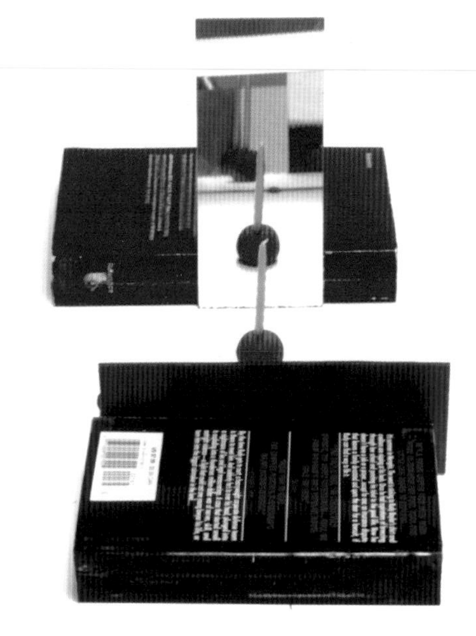

Series of reflections in two parallel mirrors.

REVERSAL OF FORTUNE

Leonardo Da Vinci was born in 1452, in Tuscany, Italy. He is best known for his famous painting, the Mona Lisa. He was a painter and also a respected engineer and scientist. He studied human anatomy, experimented with color and light, and designed mechanical devices, including flying machines. His designs did not always work, but he predicted many modern-day inventions, such as an underwater diving suit, airplanes, and helicopters. Da Vinci wrote journals in mirror writing. There are many theories about why he did this. Some people think he did this to keep his work a secret, while others believe he was dyslexic and naturally wrote backwards. See if you can experiment with mirror writing.

Hold the book up to a mirror to read the message.

You Will Need

carbon paper*

paper and pencil

mirror

book or wall

a friend

**You can make your own "carbon" paper by rubbing a thick layer of pencil or dark crayon on a regular piece of paper.*

What to Do

1. Place the carbon paper, carbon side up, under a sheet of paper.

2. Write or print your name, a sentence, or even your address on the sheet of paper.

3. Turn the paper over to reveal the reversed writing. Ask a friend to try to read what is on the paper.

4. To decode this mysterious writing, simply prop a mirror up against a wall or book. Hold your paper up with the reversed writing close to the mirror, so that it reflects the image of the writing. Can you read it now?

5. Try reverse-writing the letters with your pencil, without using the carbon paper, and view it in a mirror. How successful was your reverse writing?

6. Try writing words like NOON, MOM, POP, and AHA. How do they appear when reversed?

What Happened

The mirror changed the reverse writing back into a form that was easy for you to read. This is because what you see in the mirror is a reversed or "mirror" image. The light which hits the mirror is reflected back to you.

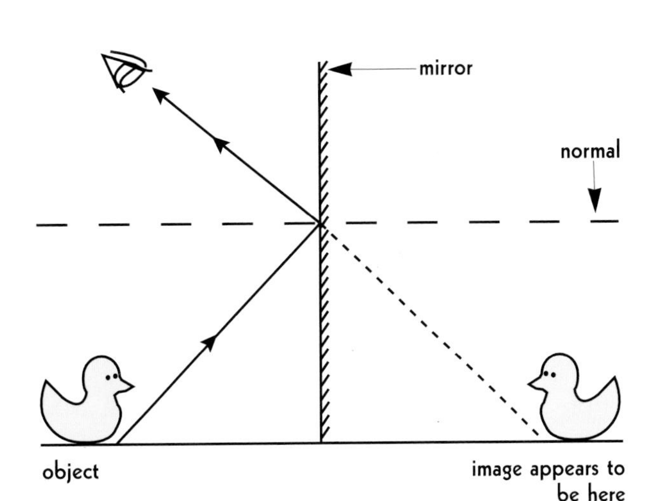

How a plane mirror reverses an image.

LOVIN' SPOONFUL

One interesting experiment to do at the dinner table involves a spoon and your nose. By breathing lightly on the spoon, you can make it stick to your nose. This brings gales of laughter from children at the table, and generally frowns from the parents. Here's a science experiment involving spoons which parents will approve of your doing, even at a fancy dinner.

You Will Need

several different types of shiny metal spoons: soup spoons, teaspoons, serving spoons

What to Do

1. Look at your reflection in the outside or back (the convex side) of a spoon.

2. Turn the spoon over to the inside or front (concave side) of the spoon, and look at your reflection in the surface.

What Happened

The surface of your spoon acted like a curved mirror and distorted your image. When you looked at your reflection on the back (convex) side of the spoon, it was smaller than you and right-side up. Straight lines such as corners of the ceiling looked curved. On the inside of the spoon, your reflection was small and upside down. If the mirror is large enough and you are close to it, your image will be right-side up and larger. It was upside down in the spoon because the spoon is quite small and you can't get near enough to the surface to see a right-side-up image.

Images formed by convex mirrors, like the back side of the spoon, are always right-side up. Images formed by concave mirrors, like the inside of the spoon, can be either right-side up or upside down, depending on the distance between the mirror and the object being reflected.

PART TWO

The Speed of Light & Refraction

What do lanterns, lasers, mirrors, and the moons of the planet Jupiter have in common? If you said that they have all been used to measure the speed of light, you would be right! The speed of light is a tricky thing to measure. Light travels much faster than anything that early scientists could measure. Light even travels faster than sound. This is why you always will see a lightning flash first before hearing thunder. The light reaches you much faster than the sound.

An Italian scientist named Galileo Galilei (1564–1642) tried to measure light's speed by seeing how long it took for light to travel from a lantern on one hilltop to an observer on another hilltop. He was unable to measure the speed, because it appeared to happen instantly—too fast for a human to record.

Light takes longer to travel from the planets to us than it does to travel from one place on Earth to another. In 1675, a Danish astronomer named Olaus Roemer measured the time it took for one of the moons of Jupiter to be eclipsed as it passed behind the planet, something that could be seen through a telescope from Earth. The time appeared to be shorter by several minutes when Jupiter was closer to the Earth in its orbit than it was when Jupiter was farther away. Roemer thought, correctly, that the difference occurred because of the extra time it took the light leaving Jupiter's moon to reach Earth when the distance between the planets was greater.

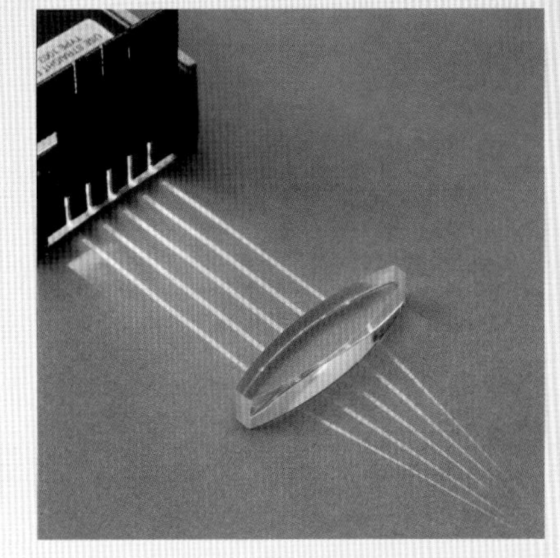

Successful measurements of the speed of light on Earth were carried out by the Frenchman Armand-Hippolyte-Louis Fizeau in 1849 and later by an American scientist named Albert A. Michelson (1852-1931). Michelson measured light traveling between a rotating many-faceted mirror and a distant mirror. The mirrors were placed on mountaintops about 21 miles (35 km) away from each other. By adjusting the speed of the turning mirror, it was possible to reflect the light back to the next facet of the turning mirror. The speed of the turning mirror was then measured mechanically and used to calculate the speed of light.

Light travels through space at a speed of 186,282 miles per second (299,792 km per second). Light travels at different speeds when it travels through other materials. When light travels from one material, like air, into another material, like water or glass, the light changes speed. If a light beam enters the second material at an angle, the light beam bends. This bending is called refraction. Lenses make use of refraction to bend light and make objects appear nearer. Some lenses concentrate light in one spot. In this section of the book, we'll do some experiments with refraction.

LIQUIDS IN LAYERS

When you sit at the side of a pool and dangle your legs in the water, do your legs look like they aren't attached to your body in the right place? Why does the water seem to change their position? Here is an experiment to show how this works.

You Will Need

clear glass or jar that can hold at least 1 cup (250 mL)

½ cup (125 mL) water

½ cup (125 mL) cooking oil

straw

What to Do

1. Pour ½ cup (125 mL) of water into the glass. Put the straw in the glass. It should rest at an angle on the rim of the glass. Look at the straw from the side.

2. Carefully pour ½ cup of oil on top of the water so it forms a layer.

3. Look at the straw from different places: above the glass, at the side of the glass, and directly in front of the glass.

Straw in air and in water.

What Happened

When the straw was placed in the glass of water, it appeared to be broken. You could see the straw because of the light reflecting off its surface. The light travels through the different mediums of air, glass, oil, and water at different speeds. When the light travels between two different materials, the change in the speed of the light at the boundary causes the light to bend. This bending of light is called refraction. Oil bends light more than water. (See the box about refraction on page 21 for more information.)

Straw in air, oil, and plain water.

Refraction

Light travels very quickly; in air or in a vacuum, about 186,000 miles (300,000 km) in a second. When light travels through other materials (for example, glass, water, or oil), it travels more slowly than it does through air. For example, light travels through water at about 140,000 miles per second (226,000 km per second). Light travels through glass at about 124,000 miles per second (200,000 km per second). Different substances slow light by different amounts, depending on their optical density. When light passes from one substance to another at an angle other than 90 degrees to the surface, the change in the speed of light at the boundary between the substances causes the light to bend. The bending of light is called refraction.

The ratio of the speed of light in a vacuum to its speed in a substance is called the index of refraction for the substance. A substance with a large index of refraction causes light to bend a great deal; a substance with a small index of refraction bends light less.

The Indexes of Refraction of Some Substances

Substance	Index
Air	1.0003
Water	1.33
Window glass	1.51
Ethanol	1.36
Cottonseed oil	1.47
Diamond	2.42

OPTICS YOU CAN EAT

Did you know that the same substance that is part of a dessert you love can actually be used in a great optics experiment? Strange as that may sound, gelatin can be used to explore refraction. Here is a simple way to perform an experiment that normally requires special equipment!

You Will Need

1 tablespoon (15 mL) powdered unflavored gelatin

mixing bowl and spoon

1/4 cup cold water

3/4 cup (250 mL) boiling water

round plastic container that holds 2 cups (500 mL)

sharp knife

ray box with flashlight*

baffle with one slit*

cookie sheet

spatula

ruler

2 transparent plastic protractors

pencil and paper

*See the Light in a Box project.

What to Do

1. Sprinkle the gelatin over the cold water in the mixing bowl and let it soak for 3 minutes to soften it. Once the gelatin is softened, have an adult add the water that just boiled, and stir to completely dissolve. Do this in a mixing bowl and not the container, as you do not want any impurities in the final product.

2. Carefully pour the mixture into a clean, round 2-cup (500 mL) plastic container. Place the container in the refrigerator for several hours, until the gelatin is set. Try not to disturb the container while it is setting.

3. Remove the set gelatin from the container by turning it upside down over a cookie sheet. If it doesn't want to pop out, place the container right-side up in a bowl of warm water for several seconds; then invert the container again over the cookie sheet. The gelatin should slide out easily.

4. Have an adult use a sharp knife to cut the gelatin in half to form two semicircular blocks.

5. Attach the baffle with one slit to the ray box.

6. Mark a large cross in the center of the white cardboard or paper. Place two protractors with their straight sides together on the upright line of the cross.

Gelatin cut in half before placing on protractor.

7. Use a spatula to lift one of the gelatin blocks and place it on top of the far protractor so that the curve of the gelatin block aligns with the curve of the protractor.

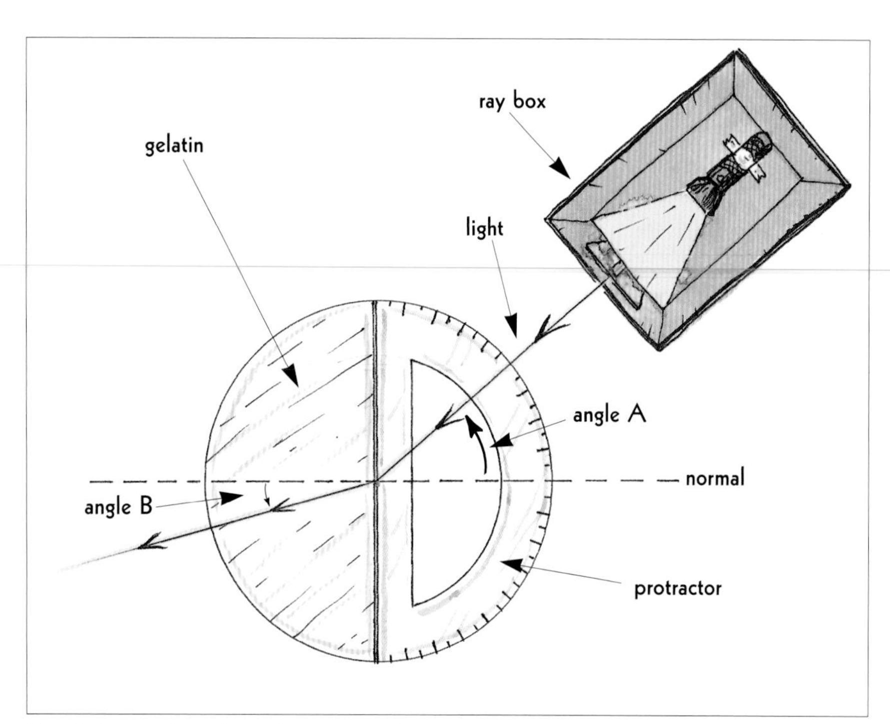

The light ray hits the gelatin at angle A but is bent towards the normal (angle B).

Photo of ray box setup and gelatin on one of the protractors.

8. Align the ray box so the light hits the gelatin at an angle from the normal when the flashlight is turned on. (The normal is a line perpendicular to the straight edge of the gelatin.)

9. Turn off the room lights. Turn on the flashlight and look at the light passing through the gelatin. Looking at the protractor under the gelatin, note the angle the light is making with the normal in the gelatin (in our diagram it's angle B). Record your findings.

What Happened

The angle the light made with the normal going into the gelatin was greater than the angle the light made with the normal as it passed through the gelatin. We say the light bent "towards the normal" because the light bent closer to the line that is perpendicular to the gelatin. When light passes from a less dense medium (air) to a more dense medium, such as gelatin, it bends towards the normal. The only light beam that wouldn't be bent is one going exactly perpendicular to the gelatin (along the normal).

COME TOGETHER NOW

Lenses are curved pieces of a transparent material such as glass or plastic that can refract or bend the light that enters them to form an image. Lenses are not a new invention. The earliest lenses, which were known to the Greeks and Romans, were water-filled glass spheres called burning glasses, used to magnify things and start fires. In the 11th century, the Arab scientist Alhazen described the magnifying power of a lens. In the late 1200s, people in China and Europe already were wearing eyeglasses made with glass lenses. Today, lenses are part of eyeglasses, cameras, microscopes, telescopes, projectors, and many other devices. There are two main types of lenses: concave lenses, which are thicker on the edges and thinner in the middle, and convex (converging) lenses, which are thinner at the edges and thicker in the middle. The word "lens" comes from the word "lentil," because convex lenses are lentil-shaped. The next experiment looks at the effect of lenses on light.

You Will Need

ray box with flashlight*

baffle with 3 to 5 slits*

paper

convex lens, or drinking glass filled with water

pen or pencil

concave lens

See the Light in a Box project for ray box and baffle instructions.

What to Do

1. In a darkened room, place the ray box at one edge of a piece of paper, aligned so that the parallel light beams from the flashlight shine across the paper.

2. Place a convex lens or a drinking glass filled with water in front of the slits in the box so the light beams from the box travel through the lens or drinking glass. Note how the light beams change direction as they move from the air through the lens or drinking glass and then out into the air. Trace the lens and beams lightly on the paper with a pen or pencil.

3. Remove the lens and the paper and replace with a concave lens and a new paper. Note the direction of the light beams as they change when moving from the air through the concave lens and out into the air again. Trace the lens and beams lightly on the paper with a pen or pencil.

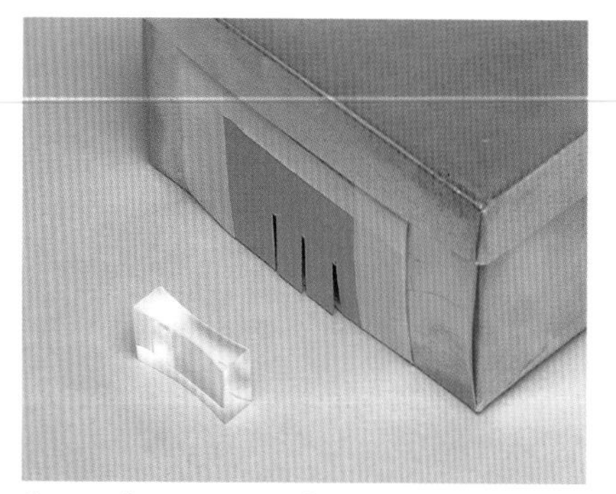

Setup for concave lens.

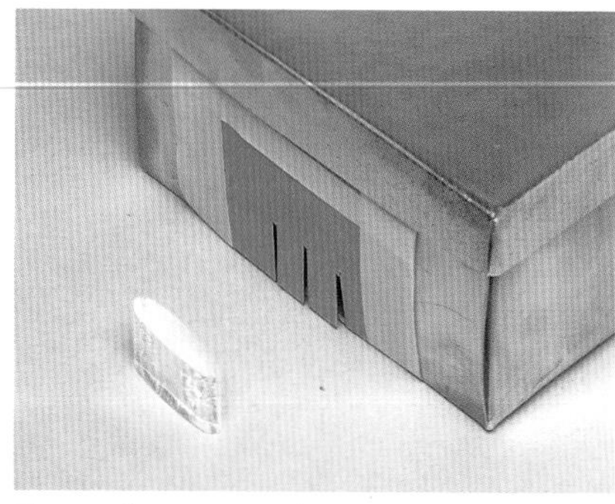

Setup for convex lens.

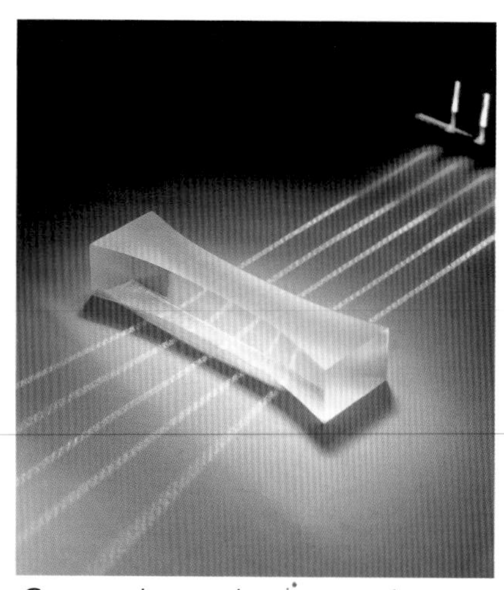

Concave lens with commercial ray box.

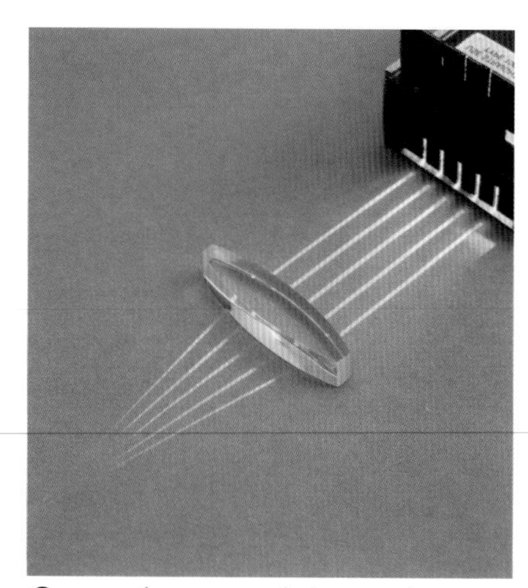

Convex (converging) lens with commercial raybox.

What Happened

You saw the light beams bend as they traveled through the lenses. The light rays moving through a convex lens from the air converge or move together. The light rays that were parallel when they came out of the ray box came closer to each other in the lens and met at one point after passing through the convex lens. When the light rays enter the concave lens, they are bent outwards; then, as they enter the air on the other side of the lens, they are bent outwards even more. The light rays that hit the lens at a large angle to the normal are bent more than the light rays that hit the lens at a smaller angle to the normal, for both lenses.

Lenses

Convex (converging) and concave (diverging) lenses come in a variety of shapes. Here are some typical ones.

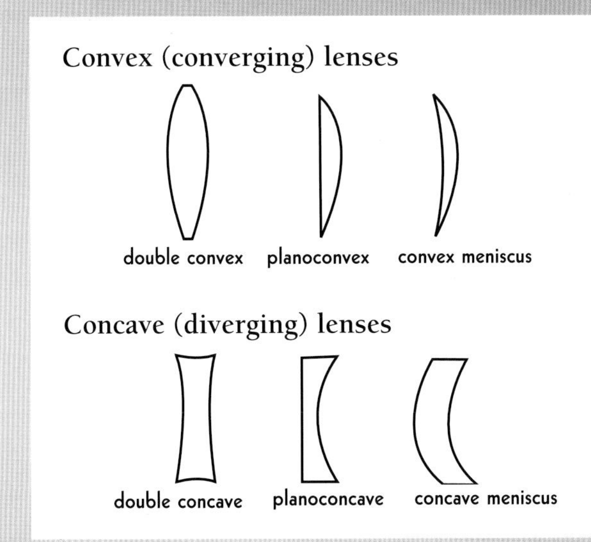

The whole world looks different when seen through a lens. Lenses make use of refraction to bend light. Depending on the shape of the lens and the distance from the object, this can have different effects. Let's look at how this happens. You can see an object because it reflects light. When the reflected rays of light pass through a convex (converging) lens, they are bent inward. If the original rays were parallel, they pass through one place, where an image is formed. The place is called the principal focus of the lens (F). Its distance from the center of the lens is called the focal length.

When an object is placed between 1 and 2 focal lengths from a converging lens, the viewer sees a real, inverted or upside down image that is larger than the object. When an object is placed closer to the convex lens than the focal point, the light rays bend outward. Your eye looks back along these diverging rays and sees a virtual image. It is upright and larger than the object. An object placed more than 2 focal lengths away from the lens forms a smaller, real, inverted image.

When parallel rays pass through a concave (diverging) lens, the rays spread out. Because the eye sees light as if it travels in straight lines, a person looking through a concave lens always sees an image that is smaller than the object and upright. It is a virtual image.

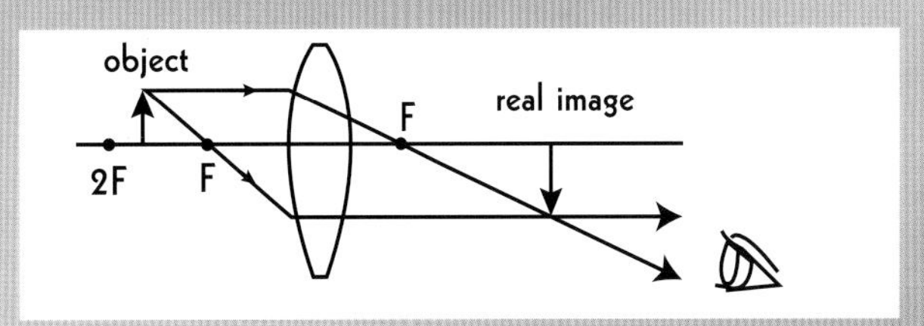

1. How a real image is formed by a convex (converging) lens.

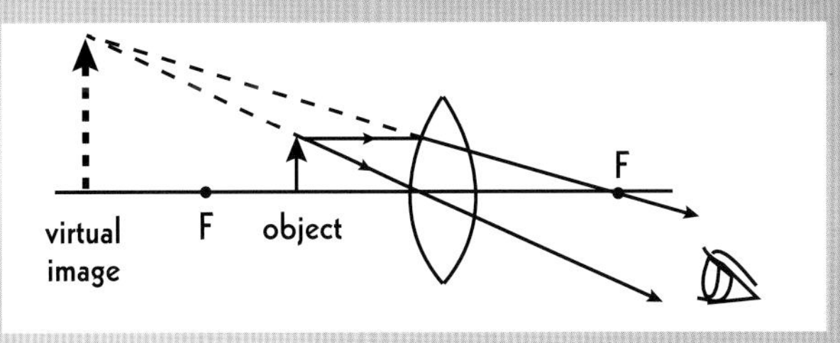

2. How a virtual image is formed by a convex (converging) lens.

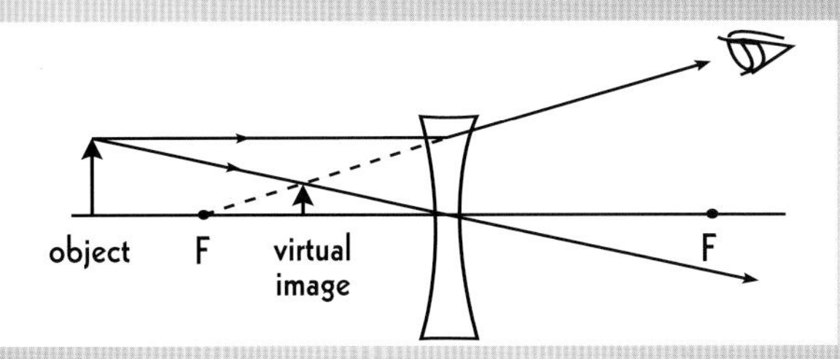

3. How a virtual image is formed by a concave (diverging) lens.

RIGHT SIDE UP...UPSIDE DOWN

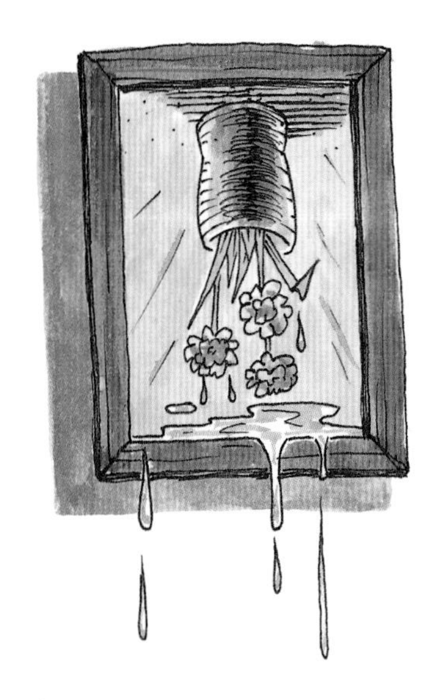

When the print in a book is too small to read, you could find a photocopier to enlarge it or use a magnifying glass. Magnifying glasses are convex lenses. They produce an enlarged image of the object you are looking at. This image grows larger if you move the lens a little distance away from the object and closer to you. Other positions of this lens can give some interesting results!

You Will Need

convex lens

printed page of paper

concave lens

What to Do

1. Place the convex lens on top of a printed piece of paper. Note the size and direction of the letters.

2. Slowly move the lens away from the paper. What happens to the letters?

3. Continue moving the lens away from the paper until the letters disappear and then reappear. What do the letters look like now?

4. Move the lens even farther away from the letters. What happens now?

5. Try this with a concave lens.

What Happened

When the convex lens was placed directly on the paper, the letters looked almost the same size as they were without the lens. As the lens was moved away from the paper, the letters looked larger and right-side up; this image of the letters is a virtual image (see glossary). As you moved the lens closer to your eye and away from the paper, the image became blurry. Then an image appeared, only this time the image was upside down and slightly magnified. This upside-down image is a real image (see glossary). Gradually the real image became smaller as the lens was moved farther from the print, until finally the print seen through the lens appeared smaller than the print on the page. The concave lens gives only a smaller, right-side up, virtual image of the printed page, in any location.

Virtual images: Person's right hand holds a convex lens. Person's left hand holds a concave lens.

Real image of type through convex lens.

Part Three

Light & Color

What is light? How do we see color? These are questions that always have fascinated people. In this section we'll use a prism to make a rainbow and answer that question you've always wanted to know about: Why is the sky blue?

ISAAC NEWTON AND THE SPECTRUM

One person who was curious about color was an English scientist named Isaac Newton. In 1672, he looked at a narrow beam of sunlight shining through a prism. A prism is a transparent block of glass or plastic or similar material, which is shaped like a triangle on each end. Each of its three other sides have rectangular faces. Newton observed that when light from the Sun passed through the prism, it was split up into colored lights that had the colors of a rainbow: red, orange, yellow, green, blue, indigo, and violet. Newton called this rainbow of colors a spectrum. He concluded that the sunlight was made up of light of many colors all combined, and explained that the prism separated the light into its parts.

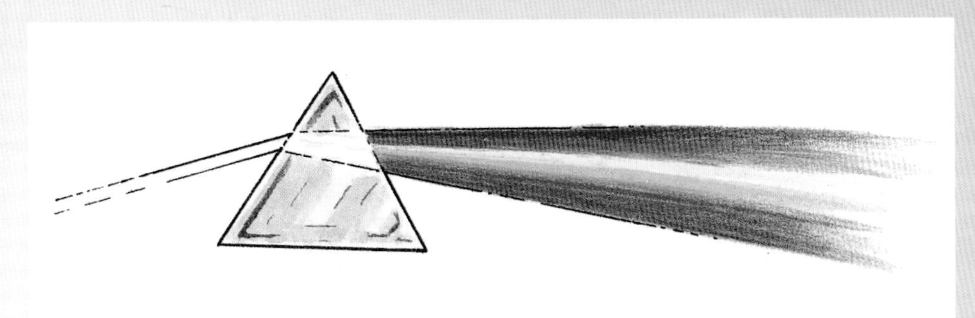

A prism breaks white light up into its component colors.

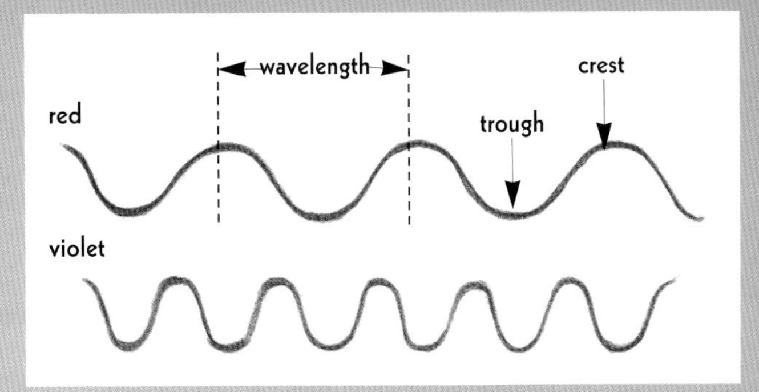

Comparison of two light waves. A light wave is measured from the same point on two adjacent waves (for example, from crest to crest). The red waves (wavelength, 700 nm) are longer than the violet waves (wavelength 400 nm), but all of them are very tiny, as a nanometer is 1/1,000,000,000 of a meter.

HOW DOES A PRISM WORK?

Light is made up of electromagnetic waves of a certain length. These waves are similar in some ways to the waves formed when an object is dropped into water. Water waves have hills and valleys, called crests and troughs. The distance between two crests is a wavelength. Visible light has wavelengths also, but they are very tiny—millions of times shorter than water waves. They are measured in nanometers or nm. A nanometer is 1/1,000,000,000 of a meter. Different colors of light have different wavelengths. For visible light, red

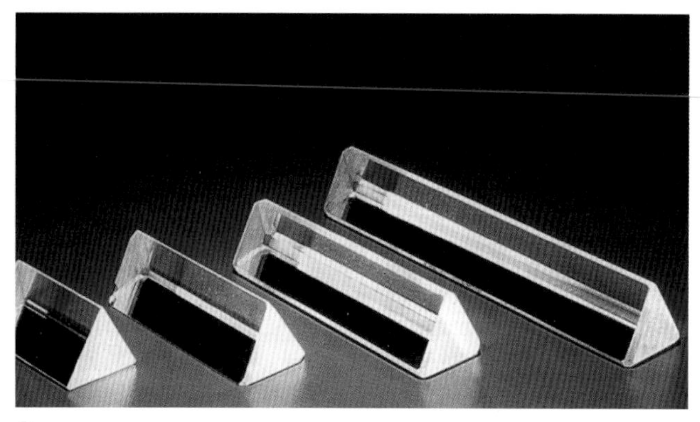

Prisms.

light has the longest wavelength (700 nm) and violet light has the shortest (400 nm). When white light goes through a prism, the different colors that make up the white light pass through the prism at different speeds. The short-wavelength light (violet) is bent more than the long-wavelength light (red). Each color bends a slightly different amount, and so the white light is split up (dispersed) into its parts.

GLAD YOU ASKED THAT!

"But I thought you said light travels in straight lines?" Yes. The motion of the light waves is up and down compared to the direction the energy is moving in. This kind of wave is called a transverse wave. When we look at a light ray, we don't see the individual waves moving up and down, just the light moving outward from the source.

Additive primaries of light.

WATER PRISMS

When you think of a prism, a triangular clear plastic or glass block generally comes to mind. Rainbows in the sky are made in natural prisms—drops of water in which light is refracted and reflected. Here's another way to create a rainbow using water and a mirror.

You Will Need

clear glass or plastic container

mirror

blob of modeling clay

piece of white cardboard

water

What to Do

1. Place a small blob of putty on the bottom of the mirror so the mirror will not slide when water is placed in the container. Rest the mirror at an angle against the side of the container, and press the modeling clay firmly into the bottom of the container (see photo and drawing).

2. Fill the container nearly to the top with water.

Mirror angled in water container.

3. Put the container on a table or floor or outside on the ground so that the sunlight hits the mirror. This may take some patience. The mirror should be facing the Sun.

4. Adjust the position of a piece of cardboard nearby until a rainbow falls on it, reflected from the mirror.

Side view of mirror setup.

Holding a cardboard so the spectrum falls on it.

What Happened

You created a water prism. The light was reflected from the mirror through a prism-shaped

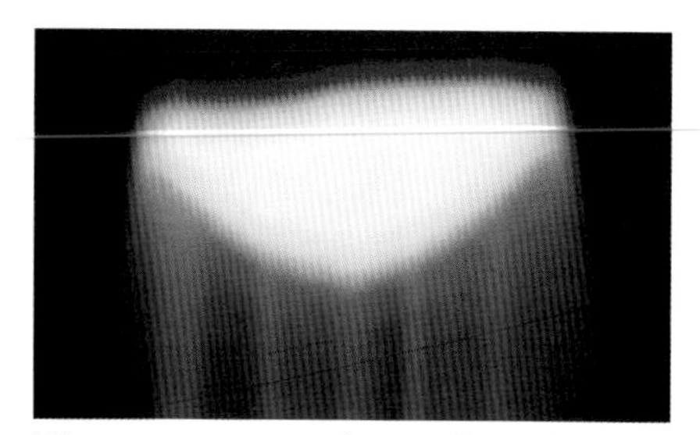
The spectrum on the cardboard.

wedge of water. This wedge of water separated the sunlight into the colors of the spectrum, as a glass prism does. You probably saw colors ranging from red to violet. One way to remember the colors of the spectrum is to remember the name Roy G. Biv. Each of the letters in this name stands for one of the colors in the spectrum: red, orange, yellow, green, blue, indigo (a bluish purple color), and violet.

What Causes Rainbows?

Where do rainbows come from? Rainbows are created when water droplets in the air act like tiny prisms and refract (bend) sunlight so its colors are separated. As with a prism, red light is bent the least and violet light the most. Some of the refracted light hits the back of the raindrop and is reflected from the back of the raindrop. Because of the angle at which you are viewing the rainbow, you see only one color of light from each droplet, so a rainbow is unique for each person who sees it.

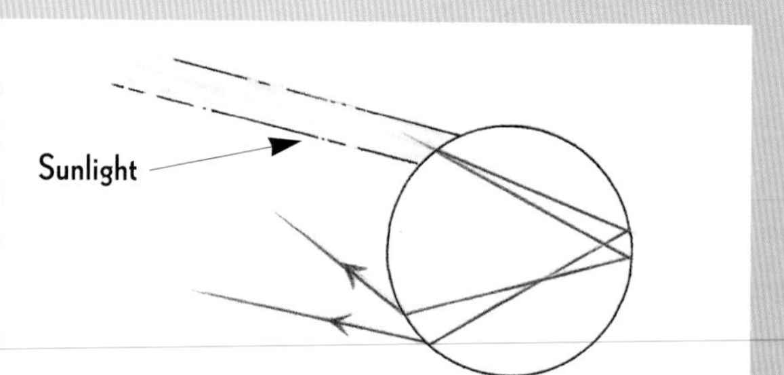

A raindrop disperses sunlight into its colors; in primary rainbows, each drop reflects light once before it leaves the drop.

There are two kinds of rainbows, primary and secondary. The primary ones have intense colors, with the red arc (curve) the highest in the sky, and the violet the lowest. The light is reflected once in each raindrop in a primary rainbow. Secondary rainbows are fainter. The light in each droplet is reflected twice in a secondary rainbow. In a secondary rainbow, the colors are reversed, and the violet arc is the highest in the sky,

Primary rainbow has red arc highest in sky.

PLASTIC RAINBOWS

When you look at a light bulb in a dark room, you only see white light, or maybe yellow light. Do you think there might be other colors there? Using a device called a diffraction grating, you can begin to discover all sorts of interesting colors around you.

You Will Need

a small piece of diffraction grating (see page 32)*

different kinds of lights: halogen, incandescent, fluorescent, indirect sunlight

white paper

colored pencils or felt makers

**Available from specialty toy and science stores or through mail order catalogues. If you can't buy a diffraction grating, you may find you have one already if you have a CD (compact disc). The back of a CD (and sometimes the front, if it isn't coated) can function as a diffraction grating if you let light bounce off it.*

What to Do

1. Hold the diffraction grating several inches (or several centimeters) away from your eyes and look at the light from a light source through the grating. Turn the grating until you begin to see rainbows on either side of the light. If you are using a CD, angle it until you see rainbows reflected off the disk.

2. Draw the color bands you see on a piece of paper. Try to make the drawing as accurate as possible. If one color band is larger or wider than another, show this on the drawing. Label your drawing with the light source.

3. Repeat steps 1 and 2 with another light source.

4. Look outside at natural light. Do not look at the Sun. Draw the color bands that you see through the grating.

5. Compare your drawings. Did the widths of the color bands change depending on the light source?

What Happened

You saw the light separated into the colors of the spectrum. Although both prisms and diffraction gratings separate light into colors, they do this in different ways. In a diffraction grating, the light moves outwards from the slits. These light waves meet waves from the nearby slits. When the peaks of the waves of the same color meet together, they make a more intense color than they would separately. This is called constructive interference. When the peak of a wave of a color meets with a trough of a wave of the same color, they neutralize each other; the result is a dark line of no color. This is called destructive interference. The surface of the CD has closely spaced spiral grooves. It behaves like a diffraction grating, dispersing white light into a spectrum of colors.

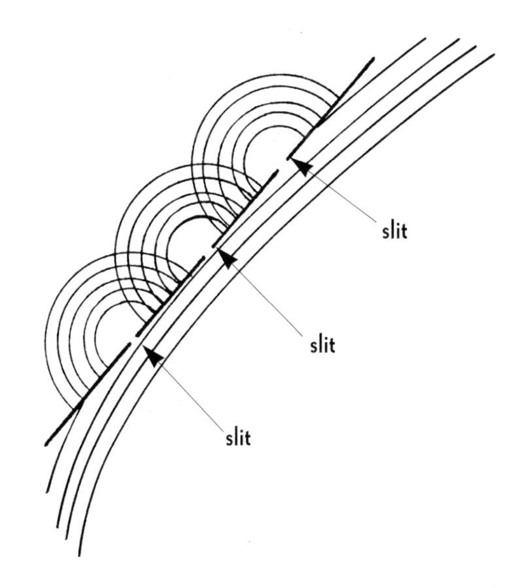

Diagram of light waves passing through slits in a diffraction grating and forming wavelets that interfere with each other.

What Is a Diffraction Grating?

A diffraction grating is a tool used for separating light into spectra (plural of spectrum) by diffraction and interference.

Diffraction is the spreading or bending of light, which can occur:

1. When light passes the edges of things

2. When light passes through narrow slits

3. When light is reflected from a shiny ruled surface.

Diffraction from slits produces fringes of parallel light-and-dark or colored bands.

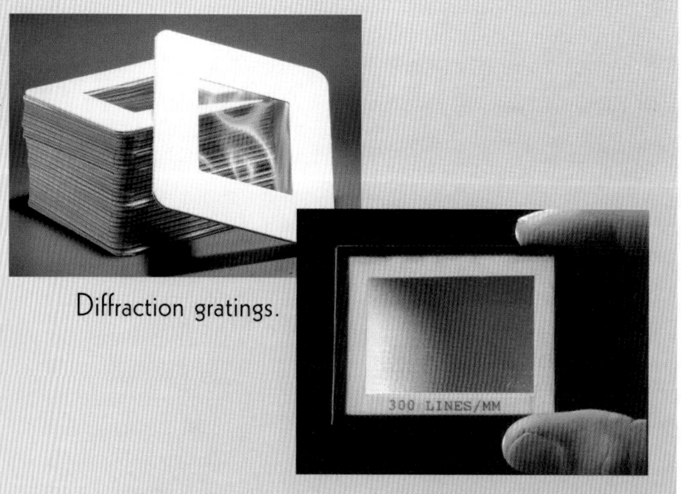

Diffraction gratings.

RAINBOW BUBBLES

Have you ever noticed how there are beautiful colors in a bubble? Now where do you suppose those colors come from when the liquid you use to make the bubble is colorless? If you closely observe a soap bubble, you will be able to predict when it will pop by watching the color changes on the surface of the bubble. Grab your favorite bubble blower and learn about the science of bubbles and color.

You Will Need

½ cup (125 mL) dishwashing liquid

several drops of glycerine (available at grocery or drug stores)

small bowl

2 cups (500 mL) water

countertop or cookie sheet

bubble blower

stopwatch or watch with second hand

What to Do

1. Place the dishwashing liquid, glycerine, and water in the bowl and mix.

2. Wet the surface of the counter or cookie sheet with enough of the mixture to cover it.

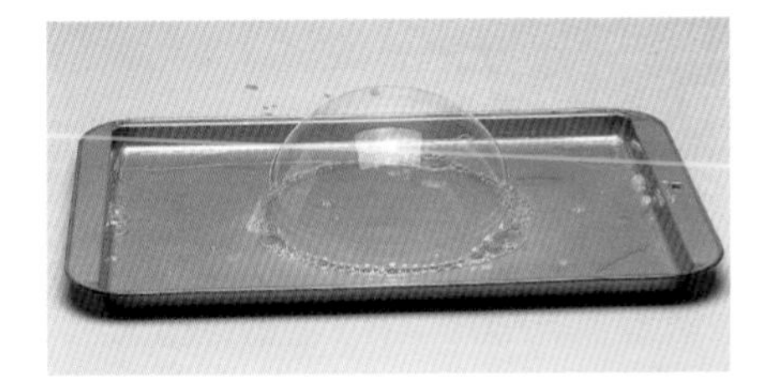

3. Dip the bubble blower into the mixture and blow a bubble onto the wet surface. The bubble will not immediately burst as long as it touches a wet surface.

4. Watch the patterns of colors on the surface of the bubble. What color does the bubble first appear? How do the colors change?

5. Blow a bubble and use a watch to measure the length of time it takes for the bubble to burst. Add several more drops of glycerine to the solution and try this again. Do the bubbles last longer with the extra glycerine? How have the colors changed?

What Happened

When the light goes from the air into the soap film, some of the white light is reflected by the inside surface of the soap film that makes up the surface of the bubble. Some is reflected by the outside surface. When light of the same color bounces off two nearby surfaces, its waves may add to each other if they get to the same place at the same time so their waves have their troughs and crests together. Then the colors are more intense. If their troughs and crests aren't together, they can interfere with each other and subtract from each other. Then there is no color or light. This is called thin-film interference, and happens with any type of material that forms a thin, transparent film. If the interference causes light waves of a certain color to cancel out, the light is no longer white and other colors are seen.

The soap bubbles did not have the same thickness over their whole surface. Depending on how thick the bubbles were, different colors interfered in different ways. As the bubble thinned, the patterns changed and you could begin to predict when the bubbles would burst. The glycerine affected the thickness of the bubbles and caused the bubbles to last longer and also to be more colorful.

PRIMARY COLORS

If you have ever mixed paints, you know that mixing red and yellow paint will make orange paint. You may have learned that red, blue, and yellow are primary colors from which all the others are mixed. If you mix them in paint, you get black. Mixing light doesn't work in the same way. The primary light colors are red, blue, and green. When you mix them together, you don't get black. Let's see what color you do get.

You Will Need

3 flashlights

cellophane: 3 pieces each of red, green and blue

3 elastic bands

screen or white paper

What to Do

1. Cover the glass of a flashlight with three pieces of red cellophane and wrap an elastic band around the end to keep the cellophane in place. Cover the glasses of flashlight #2 with green cellophane and the glass of flashlight #3 with blue cellophane in the same way.

2. Shine the light from the red flashlight onto a screen or a piece of white paper. You will see a spot of red light.

3. Arrange the green flashlight so that its light beam overlaps the light from the red flashlight. Observe the spot where they overlap.

4. Arrange the blue flashlight so that the light beam overlaps the light from the other two flashlights. Observe the spot where all three overlap.

5. Now move your hand so that it goes in between the flashlights and the screen. How does this affect the light?

What Happened

The spot where the red and green light met was yellow. When the blue light shone on this yellow spot, a white patch of light appeared. Red, green, and blue are known as additive primary colors, because all the other colors can be made by adding them together. Where the three primaries overlap you get white light. The patch of light you saw was probably not completely white, because it is difficult to find cellophane in exactly the right shades of red, green and blue.

When an object such as your hand blocks off some of the colored light, secondary colors are seen. Secondary colors are the colors that are made when two of the primary colors are combined. Yellow is a secondary color. Magenta is made from red and blue, while cyan (blue-green) is made from green and blue. Television sets and computer monitors use the additive primary colors to create the different colors of light you see on the screen.

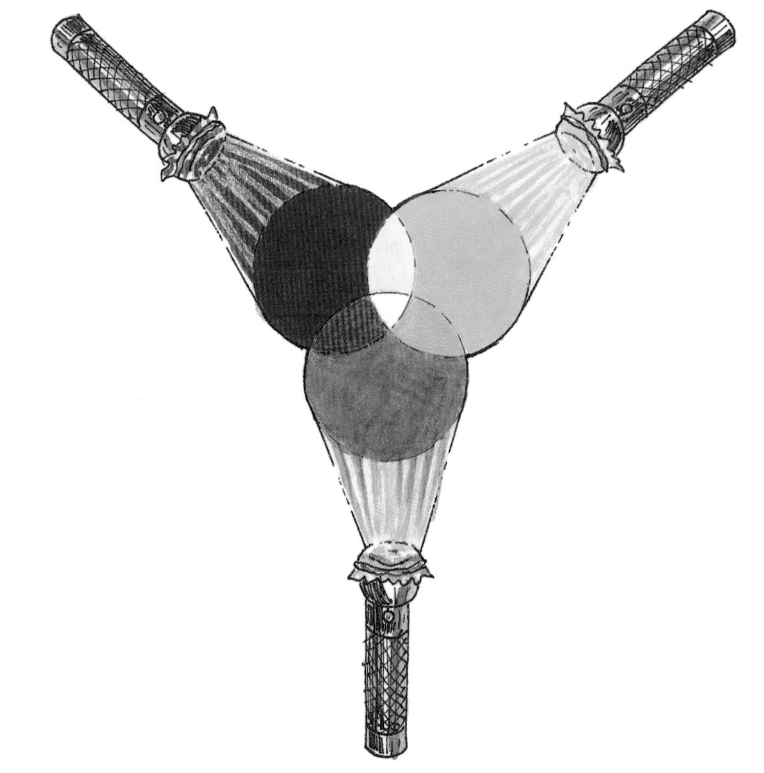

Setup of the three flashlights, showing the primary colors of light and how they form secondary colors.

Cyan

Magenta

Primary colors of light.

RAINBOW WHEELS

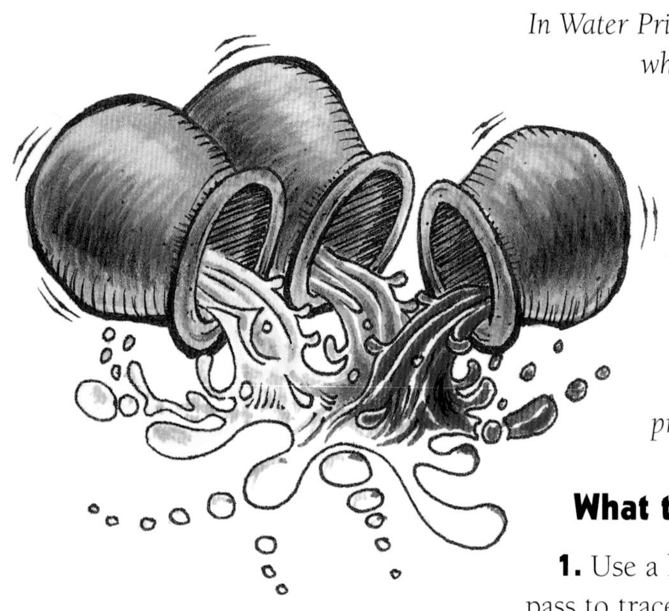

In Water Prisms, we saw how white light could be broken down into different colors. Can these colors be put back together again to form white? Here is one way to make the colors disappear and produce white.

What to Do

1. Use a large jar lid or compass to trace out a 6- or 8-inch wide (15 or 20 cm) circle on a piece of white paper. Cut out the circle and fold it in half. Carefully fold the halved circle into thirds so that when you open up the circle it is divided into six equal wedges (see diagram).

2. To make it easier for you to color you can draw lines with a pencil and ruler along the folds on the paper.

3. Glue the paper onto a piece of thin cardboard and cut the circle out of the cardboard.

You Will Need

white paper about 8 × 8 inches (20 × 20 cm)

large jar lid or drawing compass

scissors

pencil

ruler

glue

thin cardboard such as oaktag

paints or colored markers

nail

3 feet (1 m) of string

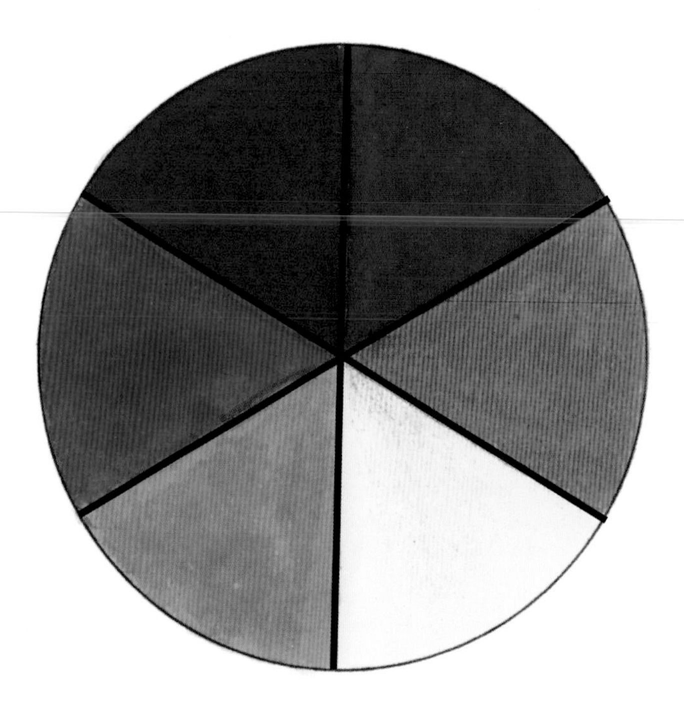

Color the wedges on your wheel as shown here.

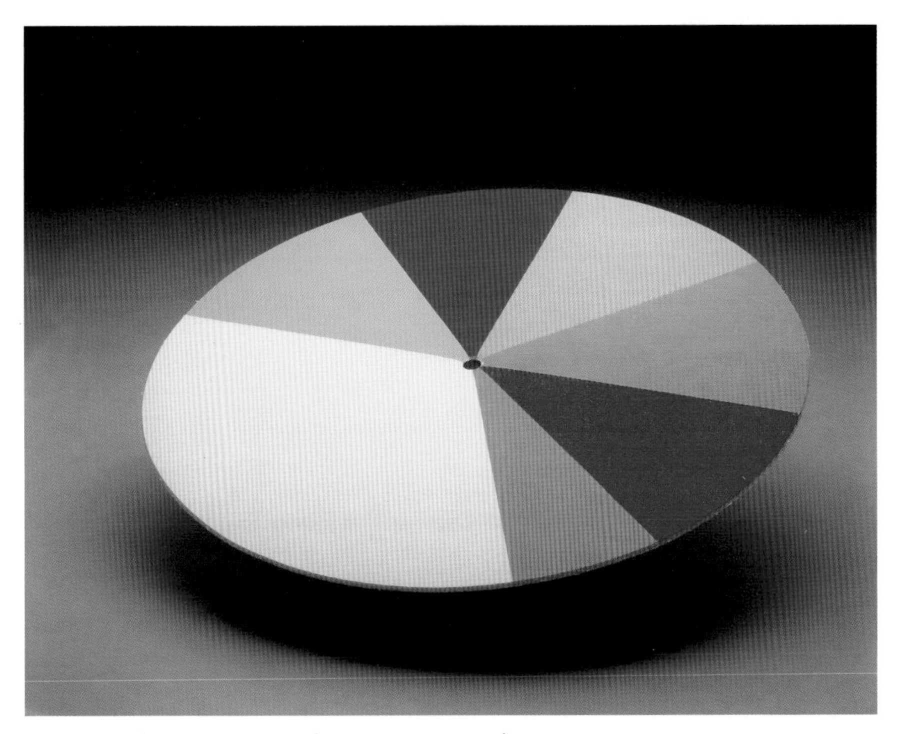

Commercial color disk produces gray tone when spun.

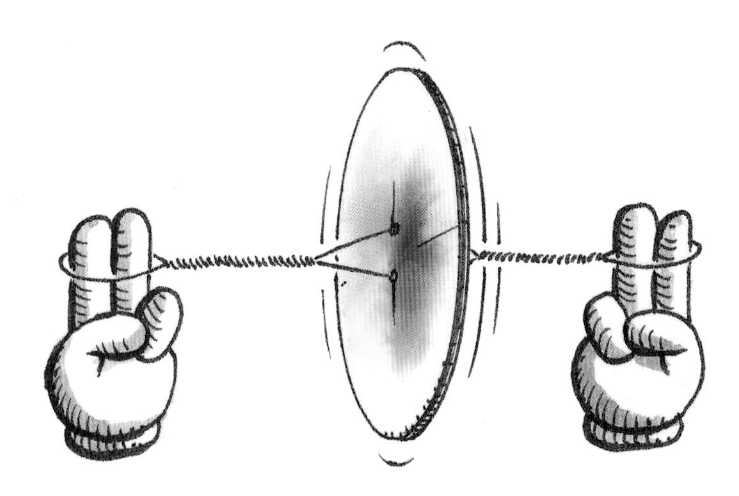

4. Paint or color the wedges in the colors and order shown on the circle on page 36. Allow the paint to dry.

5. Use a nail to poke a hole ½ inch (1 cm) from the center of the circle in the red and green wedges.

6. Thread a piece of string through the holes and tie the threads together, so that you have formed a loop. Twist each side of the loop several times; then place the ends on your fingers (see drawing). Spin the disk by pulling the string in and out. Watch what happens to the colors.

7. Create other color disks and see what happens when you spin these disks. Make a black-and-white disk, a yellow-and-blue disk, and a red-and-blue disk. What colors did you see when you spun them?

What Happened

When you looked at the spinning rainbow-colored disk, the colors disappeared and you only saw white or grey. The red-and-blue disk appeared purple when spun, and the yellow-and-blue disk appeared green. As the disk spins, the colors appear to overlap each other because the image of the sections of the wheel blend in your brain. It makes no difference if you spin the wheel clockwise or counterclockwise, the colors are still blended in the same way. If you don't spin the wheel fast enough, you will see the individual colors.

IT'S A SECRET

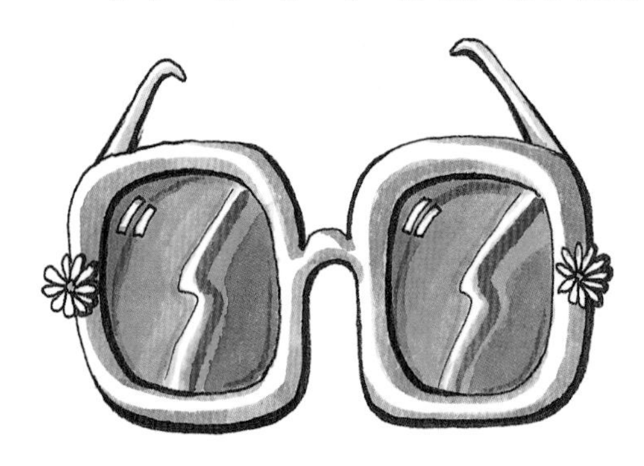

Have you ever heard the phrase "looking at the world through rose-colored glasses"? Usually this means that people see things in a positive light. What would happen if you really wore colored glasses; would you see the world differently? In Primary Colors you discovered how colors of light were created with colored cellophanes. In this experiment you will see the effect colored cellophanes have on taking away color.

You Will Need

red filter or piece of red cellophane, 2 × 2 inches (5 cm × 5 cm)

green filter or piece of green cellophane, 2 × 2 inches (5 cm × 5 cm)

this book

colored pencils or felt markers, including green, purple, and blue

paper and pencil

What to Do

1. Close one eye. Look at Figure 1 through the green cellophane or filter. Do you see any words or patterns? Write down what you see.

Figure 1.

2. Repeat Step 1, this time using the red cellophane or filter. What do you see now?

3. Create your own secret messages for friends. Write or draw whatever you want the message to be using only green, purple and blue markers or pencils. Color the area surrounding the message in any other colors you like. Remember when creating the message that the only colors which will end up looking dark are green, purple and blue when you look through the red filter.

4. Look at things in your house or outside using the red and green cellophanes. Note the effect the cellophanes have on various colored objects. Write down your findings. Visit a fabric store and look at patterned material through the cellophanes. What changes do you see now?

5. Compare the color of a green apple and a red apple when you look through the colored cellophanes. What color does each appear to be?

What Happened

The colored cellophanes made certain colors seem to disappear. The red cellophane or filter allows light at the red end of the spectrum (red, orange, yellow) to pass through, but it stops all of the cyan (blue-green) colored light and makes objects that are green, purple, and blue look dark. The green cellophane or filter allows green and a little blue or yellow light to pass through, but removes all of the magenta (purple-red) light and makes objects that are red or blue seem dark. When you look at Figure 1 through the filters, you can clearly see the hidden message.

Cyan

Magenta

WHY IS THE SKY BLUE?

When you look at sunlight, it appears to be a white light. Pictures of space usually show the space between the stars as black. Why, then, is the sky in daylight blue? What causes the sky to glow with red and orange light at sunset and sunrise? And why are some sunsets more spectacular than others?

You Will Need

large cylindrical transparent glass or plastic container, like a vase or large drinking glass

water

strong flashlight

milk (not skim milk)

white cardboard or piece of white paper

pencil and paper

What to Do

1. Fill a large glass container with water. Shine the flashlight through the water onto the cardboard or paper. Note the color of the light on the cardboard and in the water.

2. Add 1 or 2 tablespoons (15-30 mL) of milk to the container.

3. Shine the flashlight through the mixture onto the cardboard or paper.

What Happened

When the light went through the water alone, it didn't change color. When the light went through the milk-and-water mixture, the mixture appeared to be blue and the

Milk-and-water mixture appears blue.

Reddish sky at sunset is the result of Rayleigh scattering.

light that shone on the cardboard after it passed through the mixture was pinkish. The particles of fat in the milk are small and all about the same size. The light hits the fat particles in the milk and water mixture, and they scatter the light rays. The blue wavelengths of light are absorbed more easily by the particles than the red wavelengths. Then they are scattered in many directions, making the milky water appear to glow with a bluish color. The longer red light rays pass through onto the cardboard, which appears pink.

In the sky, something similar happens. During the daytime, the sky appears to be blue because the shorter wavelengths of light are more easily scattered by the air, dust, and water droplets in the Earth's atmosphere than the red wavelengths of light are. The blue light goes out in all directions and is reflected back towards us by dust and water droplets in the atmosphere. This makes the sky in sunlight appear blue.

This effect was first described by Lord John Rayleigh (1842-1919), an English scientist. Scientists call it Rayleigh scattering. We see a reddish color in the evening sky because when the Sun is low on the horizon, the light has to travel sideways through a long distance of air to get to us. The short wavelengths (blue light) in sunlight are absorbed and the longer ones (red and orange) get through. If there are many water droplets or dust particles in the air, the colors will be more intense. This is why sunsets are more vivid near the ocean, or after a volcanic eruption.

PART FOUR

Polarization

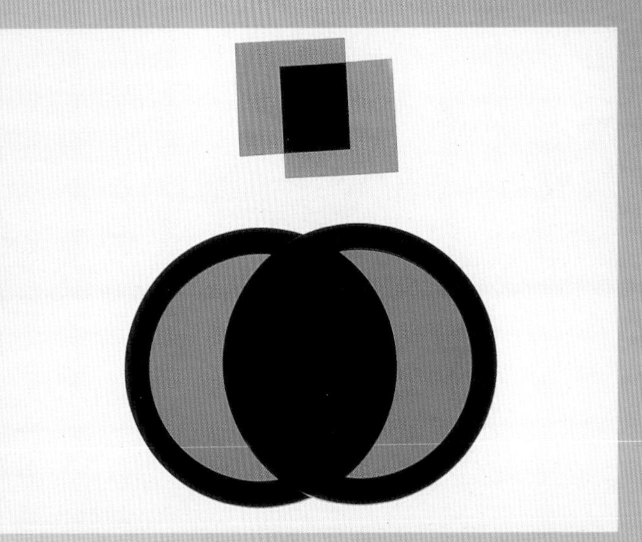

Light travels in transverse waves. Ordinarily, these waves move outwards from the source in all possible directions. Sometimes it is useful to have light in which the waves are traveling in the same direction. For this reason, people use polarizing filters, or polarizers.

Early polarizers were made of crystals such as quartz, which broke the incoming light up into two parts, one part vibrating in one direction and the other part vibrating perpendicular to the first. Part of the light could be gotten rid of, leaving polarized light. The first modern polarizing filters were invented in 1928 by a 19-year-old student at Harvard University named Edwin Land. In 1938 he invented the H-sheet polarizer. He heated and stretched a sheet of clear plastic. He coated the sheet by dipping it into a solution of iodine. Stretching the plastic changed the arrangement of the molecules of plastic so that the light coming through it was only vibrating in one plane, rather than in every direction. The rest of the light vibrations were absorbed.

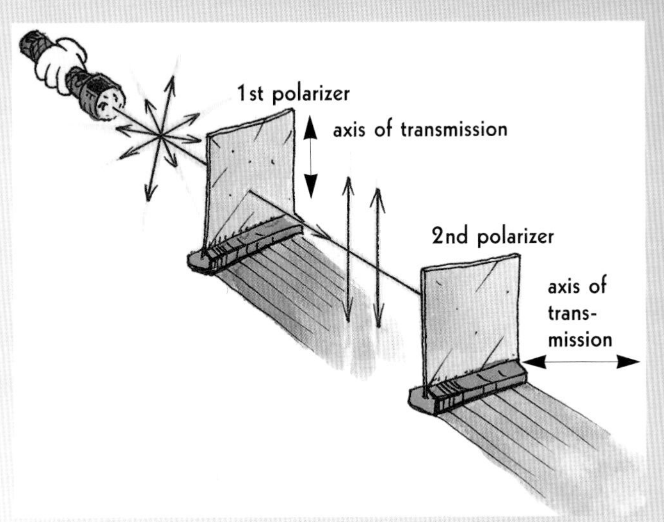

Unpolarized light and two polarizers that are at right angles to each other. Half the unpolarized light from the flashlight is stopped when it goes through the first polarizer. What passes through is polarized in one plane. The light that passed through the first polarizer is stopped by the second polarizer, if it is at right angles to the first one.

Polarizing filters are used in sunglasses to reduce glare. Sunlight is partially polarized and becomes even more polarized when it reflects off a shiny surface. The glasses absorb the light that is vibrating horizontally, but let the other light through, protecting your eyes. Polarizing filters can also be used to test lenses and stresses in plastic materials, as you'll see in the next few experiments.

Edwin Land went on to invent over 500 other things, including a camera that made instant color photos possible, which he invented for his 3-year-old daughter, who wanted to see photographs he took right away. This was called the Polaroid Land Camera.

COOL SHADES

Sunglasses are wonderful things. They make you look really cool and protect your eyes from the rays of the Sun. Many sunglasses have plastic lenses that are made of polarizing material. This material has special properties, as we'll find out.

You Will Need

two polarizing filters or polarized sunglass lenses*

light source

wax paper

** If you can't get separate polarizing filters, buy an inexpensive pair of polarized sunglasses and ask a grownup to take them apart for you so the lenses come out of the frame. You can use the lenses for this experiment and the next few about polarization.*

What to Do

1. Hold up one polarizing filter and look through it at the light. Reminder: do not look directly at the Sun, even through a filter.

2. Place the second polarizing filter in front of the first and look through both of them at the light.

3. Turn the second filter so that it is at right angles to the first filter (rotated 90 degrees). Look through both of the filters at the light.

4. If necessary, rotate the filters more, until as little light as possible passes through them.

5. Place a piece of wax paper between the polarizing filters. Look through the filters and paper at the light.

What Happened

When you looked through one polarizing filter at the light, you saw that some of the light came through the filter. When you looked through both, as you turned the filters, it became harder to see through them, until no light could shine through at all. Natural light is made up of light waves vibrating in many different directions. Using one polarizing filter screened out the light vibrating in some directions, so only about half of the light waves, all vibrating in one plane, could move through the filter. When the second filter was added and rotated, the

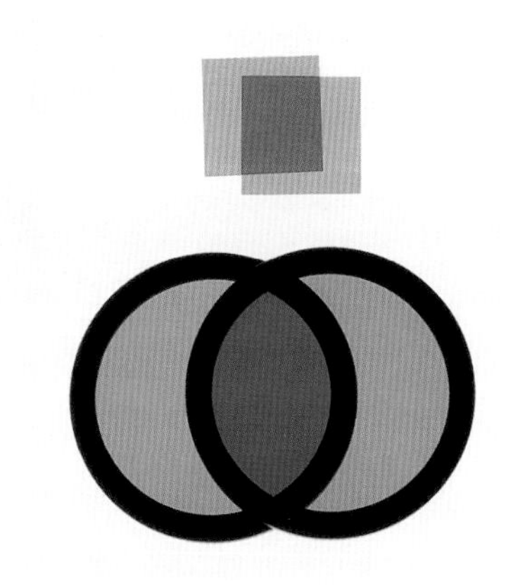

Polarizing filters aligned in the same direction.

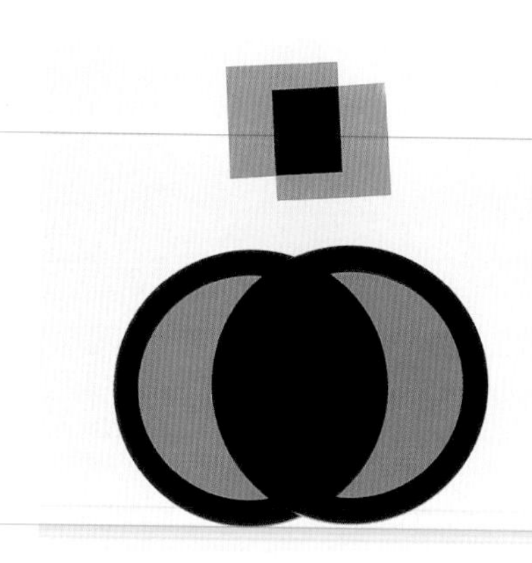

Polarizing filters at right angles ("crossed polarizers") block out light.

Crossed polarizing filters with wax paper between them.

remainder of the light passing through was blocked out, if the filter was aligned correctly. When the polarized light waves passing through the first filter hit the wax paper, they were refracted and changed directions. The change in direction of the light waves allowed some of the light to pass through the second filter.

On a sunny day, look through a polarizing filter at the light reflecting off the surface of a body of water or a puddle. The light reflected is mostly polarized in one direction. Turn a sunglass lens or polarizing filter and watch what happens to the glare.

RAINBOW IN YOUR POCKET

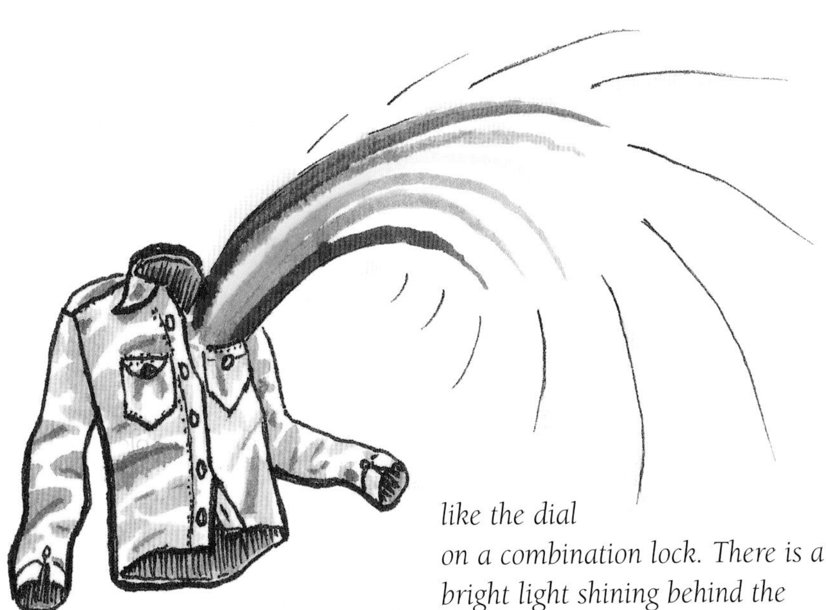

In certain art galleries and museums, you can find pictures that use a scientific principle to create changing colors. These pictures have two large glass disks mounted so that the top glass disk spins like the dial on a combination lock. There is a bright light shining behind the bottom disk. As the top disk rotates, shapes pressed between the two disks turn from green to yellow, from blue to purple, or red to orange. After you have tried this next experiment, you will see how these colors are created.

You Will Need

two polarizing filters or polarized sunglass lenses

light source

small piece of clear cellophane

microscope slide

cellophane tape (cheapest brand possible, not Scotch tape)

What to Do

1. Hold up one polarizing filter and look through it at the light.

2. Place the second polarizing filter in front of the first and look through both of them at the light. Rotate the second filter so that as little light as possible passes through the filters.

3. Place a crumpled piece of cellophane between the polarizing filters. Look through the filters and cellophane at the light. Turn the second filter and look at the change in the color of the cellophane.

Crumpled cellophane between crossed polarizing filters.

4. Remove the cellophane and place a microscope slide with a piece of cellophane tape taped to its surface between the filters. Rotate the front filter slowly and look at the tape through the filter.

Cellophane tape on a glass slide between crossed polarizing filters.

Cellophane tape on a glass slide between two polarizing filters aligned in the same direction.

What Happened

Just as in Cool Shades, when the polarizing filters were perpendicular to each other, no light came through. This position of the filters is sometimes called crossed polarizers. When the crumpled cellophane sheet is placed between the crossed polarizers, not only does some light shine through the second polarizer, but the light is colored. This happens because cellophane is doubly refractive or birefringent. When light travels through a birefringent material, it is bent in two different directions.

The light coming through the first filter was partially polarized. The light traveling through the cellophane was bent or refracted in two different directions. The refraction also split the light beams into their component colors. These colored light beams passed through the second filter, which screened out some of the colors. Only a few colors shone through. As the second filter is turned, other colors are screened out. Because the crumpled cellophane is not the same thickness all over, a number of different colors are seen.

The tape on the slide causes the same effect. As you change the amount of light passing through by turning one of the filters, the color of the tape appears to change. If you use a plastic microscope slide instead of a glass one, the whole slide will be colored. Patterns can be made on the slide using different thicknesses or layers of the tape, which will appear to have different colors when the slide is placed between the polarizers. This is how the devices in art galleries and museums are made.

STRESSED OUT

What do car windows and plastic forks have in common? Here is a seemingly silly science experiment which has useful applications in the strangest of places. You will begin to understand how science relates to everyday life.

You Will Need

two polarizing filters or polarized sunglass lenses

light source

clear plastic fork

What to Do

1. Hold up one polarizing filter and look through it at the light.

2. Place the second polarizing filter in front of the first and look through both of them at the light. Turn the front filter so that as little light as possible passes through the filters.

3. Without turning the filters, place a clear plastic fork between the polarizing filters and look through the filters at the light. Bend the tines of the fork towards each other as you are looking through the filters.

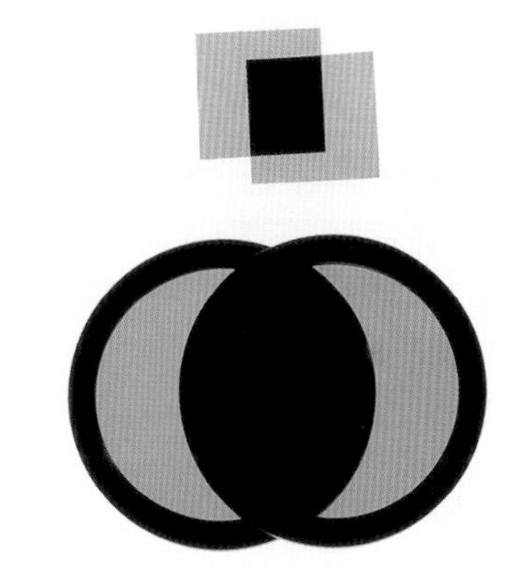

Polarizing filters aligned at right angles ("crossed polarizers").

Plastic fork between crossed polarizing filters.

What Happened

Just as in Cool Shades and Rainbow in Your Pocket, when the polarizing filters were crossed (perpendicular to each other), no light came through. When the fork was placed between the filters, you saw colors, just as with the cellophane in the previous experiment. The clear plastic used to make the fork is birefringent. When you squeezed the tines together, the colors became more intense. This is a special type of birefringence called stress birefringence. It happens when certain materials are compressed, stretched, or stressed. One example of this is the birefringence of automobile windshields. The windows are stressed during manufacturing so that they will shatter on impact. Often you can see colors where the glass curves in the windshield, if the light is shining through at the correct angle.

Part Five

Eyes & Vision

Each living creature has eyes suited to help it survive. Most birds have eyes on either side of their heads to help them hunt for food and to see predators who are hunting them. Fish have eyes that are adapted to life in the water. Insects have eyes that help them detect movement, find food or mates, and escape being eaten. People have eyes in the front of their heads, which help them see depth. In this section of the book, we'll do some experiments to explore how human eyes work.

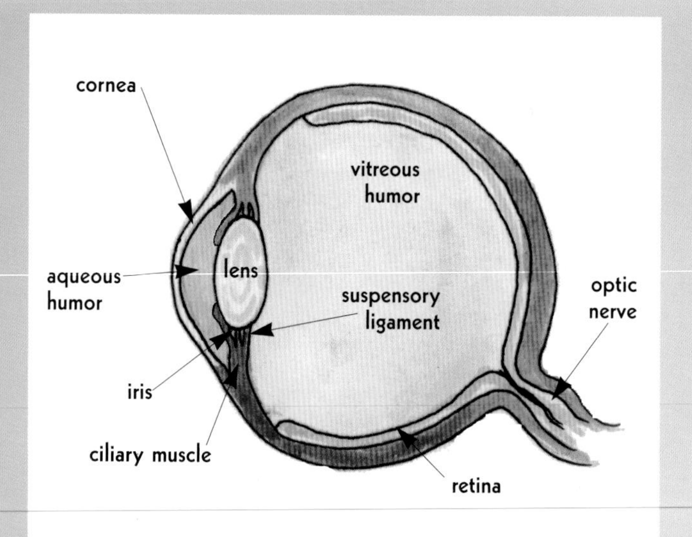

Diagram of an eye, seen in cross section.

HOW OUR EYES SEE

You don't have to think about making your eyes work; it happens automatically. Let's trace what happens when you see an image, and learn about the parts of the eye.

❑ Light enters your eye through a thick, transparent membrane called the cornea. The cornea is curved. Its index of refraction is greater than that of air, so the cornea bends the light inwards (towards the normal).

❑ Once the light has passed through the cornea, it enters a fluid-filled area at the front of the eyeball. The clear fluid is called the aqueous humor. It acts as a lubricant for the eye and helps the eyeball keep its rounded shape.

❑ The light then moves through the pupil, an opening in a muscular ring called the iris, which is in front of the lens. The iris can change shape to make the pupil larger to let in more light or smaller to let in less light.

❑ Behind the iris is the lens. The lens of the eye is a biconvex (converging) lens (fatter in the center and curved on both sides). Muscles called ciliary muscles are attached to fibers called suspensory ligaments, which attach to the outer edge of the lens. These muscles can tighten or relax to make the lens rounder or flatter, changing its focus. When the muscles are relaxed, the lens can focus on distant objects. When the muscles are tightened, the lens is fatter in the center or rounder, and can focus on close objects. The lens focuses the image (light) that passes through the cornea onto the retina. This process is called accommodation.

❑ Once the light has passed through the lens, it enters a chamber behind the lens which is filled with a jellylike substance called the vitreous humor, which acts as a support for the eyeball.

❑ The light is focused onto a light-detecting layer of cells called the retina. It includes two kinds of light-detecting cells, rods and cones. The rods function in dim light. The cone cells function in bright light. The point on the retina where vision is the sharpest is called the fovea. There is a small area on the retina called the blind spot, which lacks light-detecting cells. The blind spot is where the optic nerve, responsible for sending electrical messages from the eye to the brain, is attached to the retina. (You'll get a chance to see how the blind spot affects your vision in the Right Before Your Very Eyes project.)

❑ The eyeball itself has a thick whitish coating called the sclera, which covers the parts of the eye not covered by the cornea.

❑ Inside, between the layer of the sclera and the retina, is a layer called the choroid coat, which contains the blood supply for the eye.

❑ The image projected onto the retina by light entering the eye is an inverted or real image. You aren't aware that images are upside down on your retina because your brain has learned to turn the images you see right-side up.

Binocular Vision

How do your eyes work together? You are able to judge depth more easily because you have two eyes. Each eye sees objects at a slightly different angle. The difference in the angles is greater for nearby objects than for objects that are farther away. Your brain notes the amount of difference, and you have learned to use this to tell how far away something is. You can check this by looking at a nearby object with one eye closed; then look at the same object with the other eye closed. The image will seem to move over when you do this. It is much more difficult to judge depth if you only use one eye. Try catching a rolled up pair of socks thrown by a friend when you have both eyes open. Then close one eye and try it again. You can catch the socks with ease with both eyes open, but it's not as easy with one eye closed.

THE BIGGER TO SEE YOU WITH

Remember Little Red Riding Hood and the Big Bad Wolf? She asked the wolf why he had such big eyes. The wolf's response was really a scientific explanation of how eyes see in the dark. (That's his story and he's sticking to it.) The pupil is the dark round spot in the middle of your eye. No matter what color a person's eyes are, their pupils are always black. The pupils grow bigger and smaller, depending on the surrounding light. Here is an experiment that you can do using a great test subject: yourself!

You Will Need

mirror

light with a dimmer or adjustable switch, or a flashlight

dark room

helper

cat (optional)

What to Do

1. Stand in front of a mirror. Turn off all the lights in the room for about a minute. Have a helper turn the light or flashlight on so that it is barely lighting the room. Look in the mirror at the size of your pupils.

2. Have your helper turn on all the lights in the room, so that it becomes really bright. Look in the mirror to see if the size of your pupils has changed.

3. If you have a pet cat, you can try this experiment using the cat as a subject. Hold the cat in dim light and note the size of its pupils. Turn the lights on so the room is bright and see if the cat's pupils change.

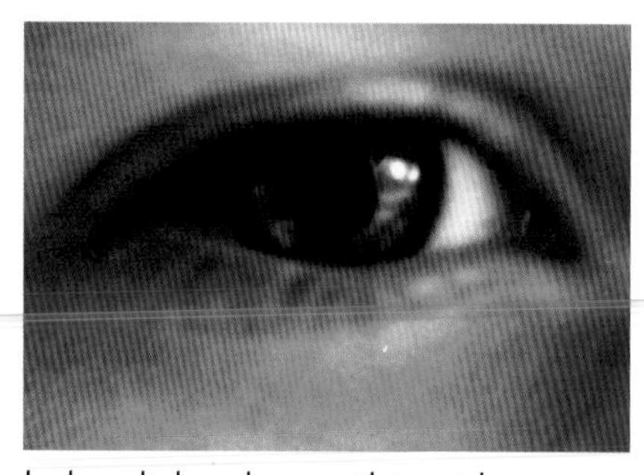

In low light, the pupil is wider.

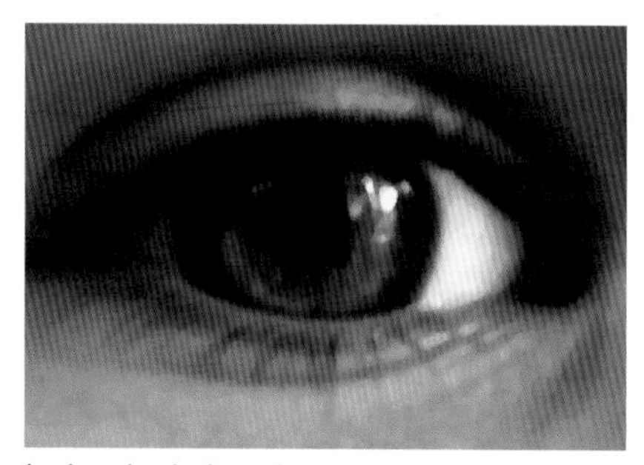

In bright light, the pupil is narrower.

What Happened

Pupils perform an important function in seeing. They allow just the right amount of light to enter your eye. When they dilate (grow wider), they are allowing in more light, making it easier to see in the dark. When they contract (grow smaller), they are really protecting your eyes from bright light. The pupil itself is just a space; it is the iris, the colored part of your eye, that does the work. The iris is a muscular ring that stretches and tightens to control the light entering your eye. The iris has layers of pigment cells that give it its color. You inherit the color of your irises from your parents.

RIGHT BEFORE YOUR VERY EYES

Do words sometimes disappear off a page when you are looking at them? This is because of something called the blind spot. You'll find out more about this in the following experiment.

You Will Need

this book

What to Do

1. In the box on this page there is a black dot and a black cross. Place your left palm over your left eye so that no light gets in.

Stare at the black dot with your right eye.

2. Pick this book up and slowly move it towards you. Keep staring at the dot while you are doing this.

3. Notice the point where the black cross seems to disappear.

4. Now do this again, this time covering your right eye with your right palm. Stare at the black cross with your left eye.

5. Pick up the book and bring it closer to you as you did in Step 2. Notice the point where the dot seems to disappear.

What Happened

You found the blind spot for each of your eyes. There's nothing to worry about; it doesn't mean you are going blind. The blind spot is the place where the retina attaches to your optic nerve. The retina is the light-detecting layer lining the back of your eyeball. It includes two types of cells that detect light. These cells send messages through the optic nerve of each eye directly to your brain. The messages are interpreted by your brain to produce the images you see. In the blind spot, there are no light-detecting cells, so no message gets to your brain when an object's image falls on that part of your retina.

WHAT'S THE POINT?

Your eyes can see objects that are nearby and also ones that are far away. This is because the lens in your eye can change its shape and its focus. Have you ever wondered why things get blurry when they come too close to your eye? This is because you have a near point. The near point is the point nearest the eye at which an object can be placed and still produce a sharp image on the retina. Let's see what your near point is and how it will probably change when you get older.

You Will Need

a table

masking tape

a pencil

a helper

a ruler

an older person

What to Do

1. Place a strip of masking tape from the edge of the table towards the middle.

2. Position your face so your right eye is level with the tabletop and at the edge of the table near the tape strip. Cover your left eye with your hand. Have a helper slowly move the tip of a pencil along the tape towards your eye. As soon as the tip of the pencil becomes blurry or out of focus, ask the helper to stop and mark the spot.

3. Measure the distance from the edge of the table to the marked spot. This distance is your near point. Measure the near point for your left eye. Is it the same?

4. Measure the near point of an older person. How does the person's near point compare to yours?

What Happened

When you measured your near point, it was probably a short distance, only a few inches or centimeters. The near point of your right eye was probably similar to the near point of your left eye. The near points of the eyes of the older person you measured probably were larger (farther away) than yours were.

When you look at something far away, the lens in your eye is relaxed and the focal distance is long. When you look at something close, the ciliary muscles tense, pulling the lens into a more curved shape. As a person gets older, the lenses in the eyes become less flexible, and it becomes more difficult to focus on nearby objects. When this happens, the person's near point gets farther away. You may find that individual people are different from this general pattern. Try testing your whole class, and have your friends test their parents and grandparents. Then compare the large group of results with each other by age of the people, and see if there is any pattern.

LINE TO YOU

Your eyes can't lie to you, or can they? Scientists have been studying perception for hundreds of years to try to learn how the brain interprets what our eyes see. Try these experiments to see what your eyes tell you. Then look at the end of the project to see the answers.

You Will Need

this book

ruler

pen and paper

What to Do

1. Look at each of the drawings on page 52. Answer the questions below the drawings. Record your answer on a piece of paper.

2. Use a ruler to measure each shape. What did you find?

What Happened

In Part I, it seemed that line AB was shorter than line CD, but actually, they are the same size. In Part II, Pyramid 1 looks bigger than Pyramid 2, but when you measured them you found they are the same size. In Part III, you saw a white triangle, even though no triangle is drawn.

Each of these drawings causes you to see an optical illusion. The images that fall on the retinas of your eyes are 2-

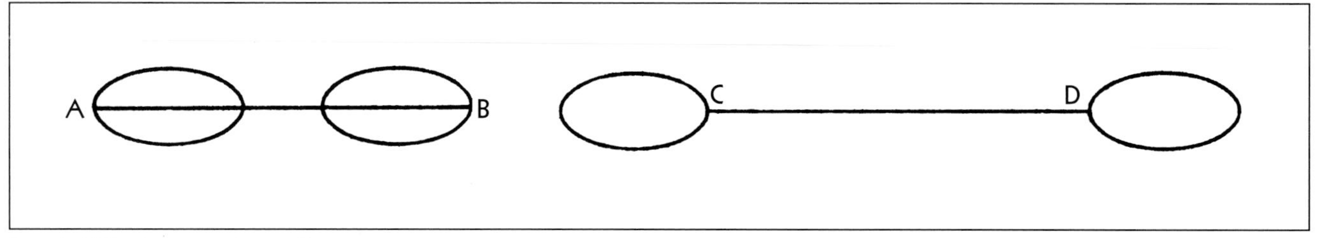

Part I. Which line is longer?

dimensional images. Your brain makes a judgment of what situation an image represents based on three factors. The first is what your eyes see. The second is the built-in pattern recognition of the brain. The third factor is your experience.

The things your eyes see are not clear or give conflicting messages to your brain, which sometimes causes the visual information to be misinterpreted. This misinterpretation creates an optical illusion. For

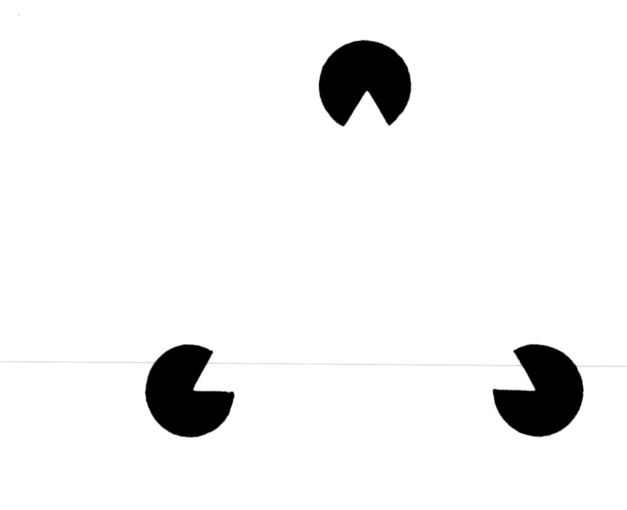
Part III. What shapes do you see?

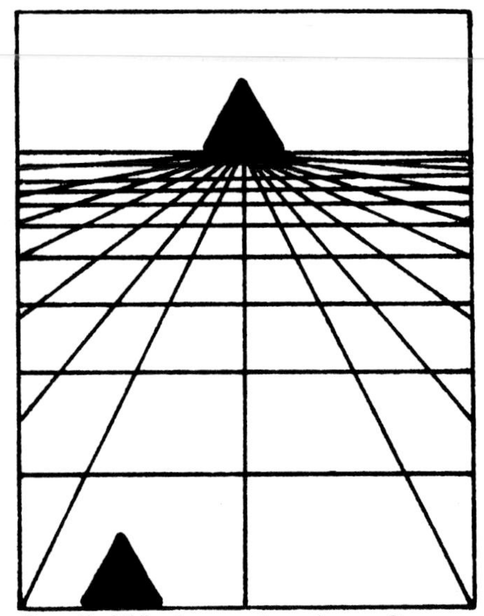
Part II. Which triangle is bigger?

example, your brain tends to see things in simple patterns. It sees a white triangle in Part III because that is the simplest interpretation of the image. In Part II, the lines coming together at the top suggest that the triangle at the top of the drawing is far away. Since it is far away, your brain estimates the top triangle's actual size as being bigger than the near (bottom) triangle. Psychologists study optical illusions to learn about how the brain puts information together.

Create your own optical illusions. Can you fool your friends and family?

BLACK & WHITE AND RED ALL OVER

What's black and white and red all over? No, it's not a penguin who is blushing. A German scientist named Gustav Fechner discovered in 1840 that when a round disk with black-and-white designs was spun quickly, a strange thing happened: he saw colors on the disk. These disks, sometimes called Benham's wheels or tops after C. E. Benham, an Englishman who popularized them, are toys found in many museum shops. You can make your own Benham's wheels and test them to see if they really work.

You Will Need

white paper

photocopier

cardboards large enough to fit the wheels on when enlarged

glue stick or paper glue

black felt-tipped marker

scissors and drawing compass

hand-held electric cake mixer or pencil

tape

adult helper

What to Do

1. Make enlarged photocopies of the designs given; between 3 and 8 inches (7.5 and 20 cm) diameter works well. If you can't use a photocopier, draw the designs, enlarged, on white paper. Then glue the designs onto the cardboard.

2. If you drew the designs, color in the designs you drew with a black marker to match the ones on this page.

3. Cut the designs out of the cardboard to create two disks.

4. Remove one beater from the hand-held electric mixer. Tape a disk to the bottom of the remaining beater (see photo). If you don't have an electric mixer, poke a hole in the center of the disk with the scissors and push the pencil through with its point down so the disk sits halfway on the pencil, to form a top.

Diagrams of wheels to make. Enlarge to between 3 and 8 inches (7.5 and 20 cm).

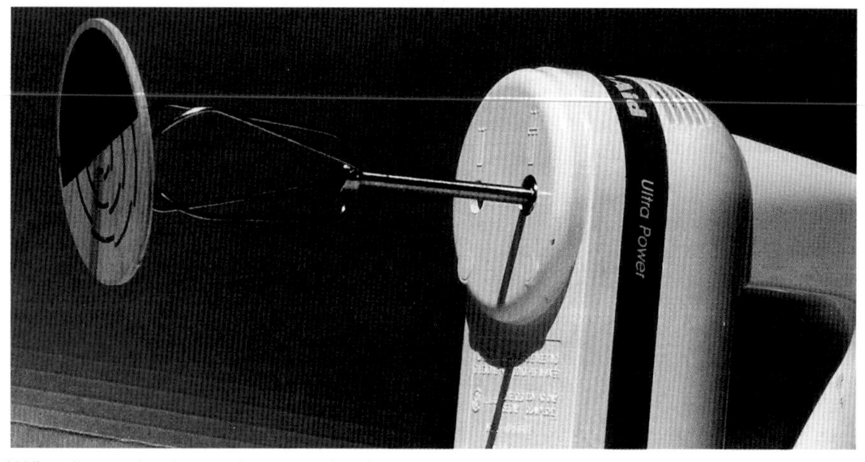

Wheel attached to a beater of cake mixer.

5. Have an adult turn on the beater and watch the disk spin. Try it at several speeds. Make sure you keep your fingers away from the beater while it is moving. What did you see? Write down your results. If you are using a pencil to spin your top, try spinning it at different speeds by hand.

6. Remove the first beater and move the disk to the other beater, so the disk will now rotate in the opposite direction when it is turned on. How does this affect what you see? Write down your results. If you are using a pencil to spin your top, try spinning it by hand in the other direction.

7. Change disks and try this again. Write down your results.

What Happened

Each disk produced different colors when it was spun. The direction that the disk was spinning, clockwise or counter-clockwise, affected the colors. It is not clearly understood why the spinning disk looks colored. Scientists think that the rapid changes between the white and black areas may trigger the area of the brain that creates the image of the colors.

THE WHOLE PICTURE

If you look at a strip of movie film, you will see a series of images separated by frames. When run through a film projector at the right speed, the separate images on the film blend together. Here's an experiment you can do to start to understand why this happens.

You Will Need

slide projector

slide

movie screen or large piece of white cardboard

2 adjoining rooms or a room and a hallway

masking tape

yardstick or metre stick or thin flat piece of wood

several friends

What to Do

1. Place the screen or cardboard about 5 feet (1.5 m) in front of a doorway between two rooms.

2. Face the projector towards the screen. If possible, the projector should be at least 6 to 8 feet (1.8 m to 2.4 m) away from the screen. Put a slide in the projector and turn on the machine. Focus the slide on the screen.

3. Mark the spot on the floor where the screen was with a piece of tape. Take the screen away and turn off the lights in the two rooms.

4. Have a friend quickly wave the stick up and down in the beam of light from the projector at the spot where the screen was. Keep the flat side of the stick towards the beam from the projector.

5. Try waving the stick at different angles or shapes, but still up and down. What happens to the picture?

What Happened

The whole image of the slide appears before your eyes. At each moment the stick moves back and forth, a portion of the image from the slide is in focus on the stick. Your eyes retain the portions for a small amount of time and your brain puts the image together. You see the whole image as if the screen were still in place. This is similar to the process that occurs when you see a series of images on your television, or in a movie. In a movie, the individual images in each frame of the film remain in your eyes and brain long enough to blend to form a whole image that seems to be moving. On a television monitor, the images of the individual dots of light remain long enough in your eyes and brain to blend to form a whole image.

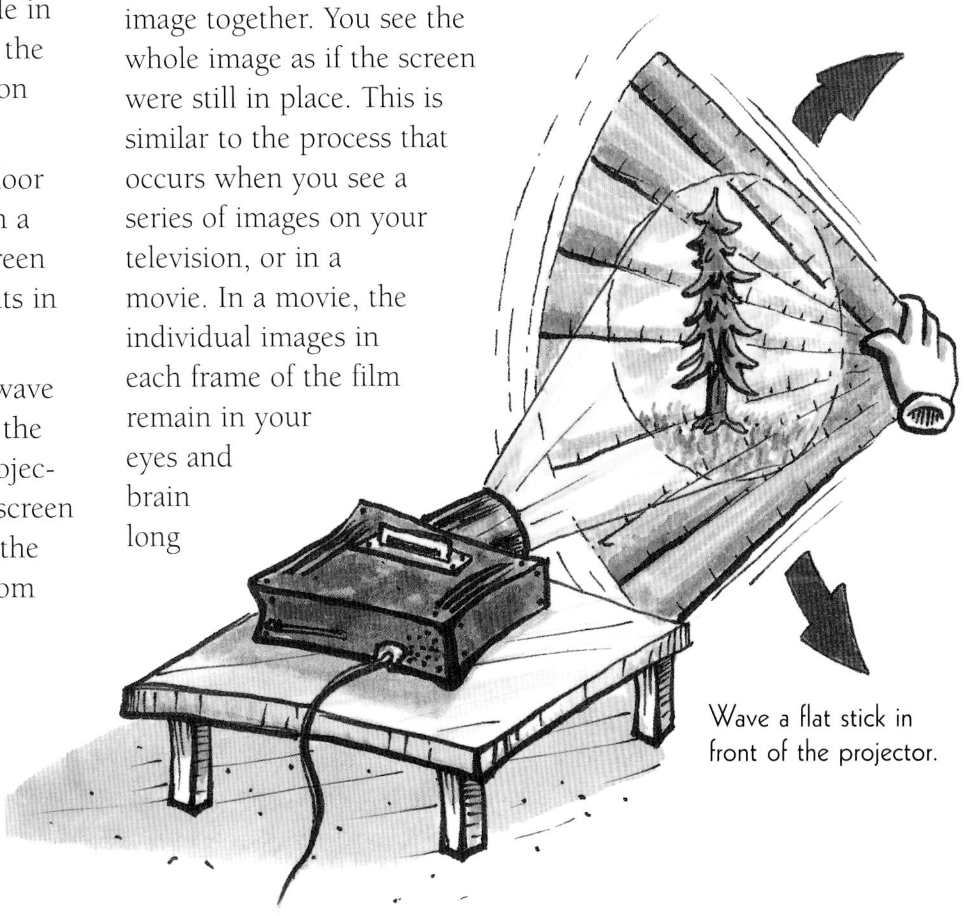

Wave a flat stick in front of the projector.

DISCO TECH LIGHTS

You Will Need

this book

drawing compass

pen and pencil

paper

cardboard

scissors and craft knife

gluestick or paper glue

fluorescent light

In some restaurants or nightclubs, there are special flashing lights that shine on the dance floor. This creates a strobe effect, which makes it look as if people are dancing in slow motion. It looks like this because you only see the dancers when the light shines on them. The flashing light gives them the appearance of moving and stopping, moving and stopping, instead of continually moving. Here is a way to see even faster flashes of light.

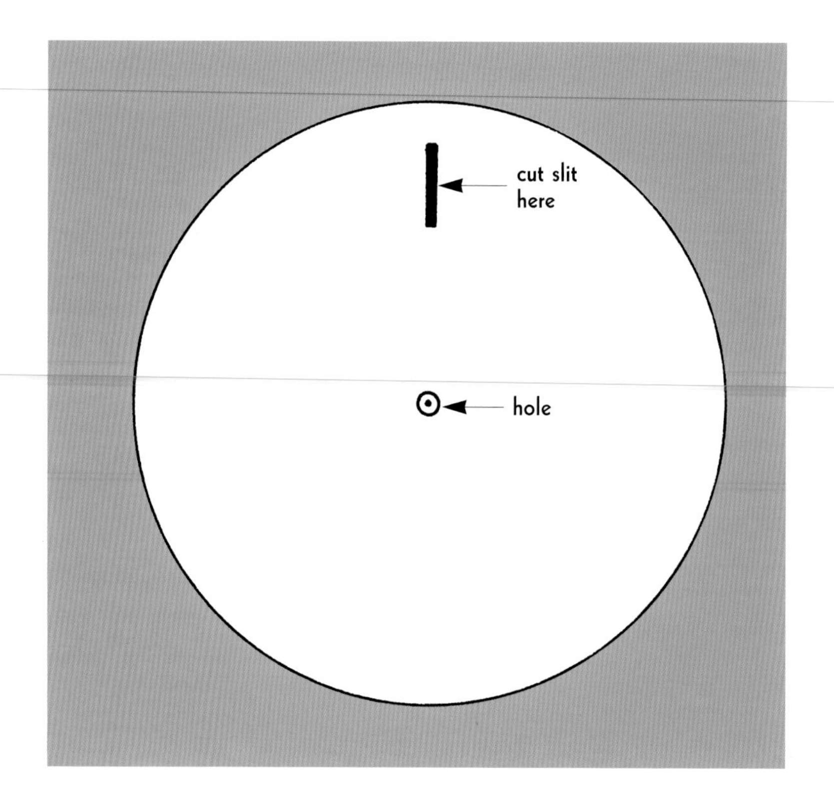

What to Do

1. With a compass, draw a 4 to 6 inch (10 to 15 cm) circle like the pattern in this book on a piece of paper; then glue the paper onto a piece of cardboard.

2. Cut out the circle. Ask an adult to cut a small hole in the center with the craft knife to fit the tip of the pencil, and draw and cut out a slit about 1 inch × ⅛ inch (2.5 × .3 cm), and about ½ inch (1 cm) in from the edge of the disk (the dark rectangle in the drawing).

3. Poke a pencil through the hole in the center. Hold your hands on either side of the pencil and rub them back and forth to twist the pencil, as if you were warming your hands on a cold day.

4. Hold the wheel in front of a fluorescent light and look through the slit at the lights. If possible, turn off all the lights in the room, except the fluorescent light.

5. Try spinning the wheel at different speeds, while looking at the light.

What Happened

You have created a strobelike effect. ("Strobe" is short for "stroboscope," an instrument used for looking at motion in such a way that things seem to be slowed down.) In the US and Canada, fluorescent lights dim and brighten 120 times per second. This is because the current of the household electricity supply changes direction 60 times each second. When you looked through the slot at the lights while spinning the wheel, the lights appeared to flash on and off because you only saw them for part of the time. In order to see this effect, the wheel has to spin at a speed that is in time with the lights' flashing.

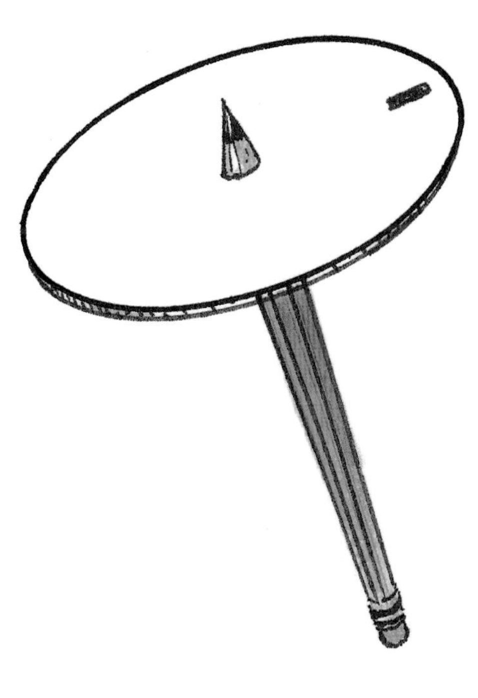

In some countries, the electricity operates slightly differently, and the lights will flash at a different speed. The flashes should be the same as twice the number of hertz or cycles per second of your electrical current. You "isolate" the cycles of light with the strobe effect, so it looks as if the light is flashing off and on.

BLUE DOG

Cajun artist George Rodrique, who lives in Louisiana, is famous for his pictures of Blue Dog. Blue Dog also could be called Yellow Dog. To find out why, try this experiment.

You Will Need

several pieces of white paper

the dog outline picture in this book

pencil and crayons

What to Do

1. Stare at the blue dog outline picture on this page for a minute. Then look at a white sheet of paper. What do you see?

Stare at this blue dog.

2. Draw a simple picture on a piece of paper. Color it all one color. If you want to see a green tree, color the tree magenta. If you want to see an red-colored fruit, color it cyan (blue-green). To see a yellow object, try coloring it blue. When you are finished with your drawing, stare at it for several minutes. Then look at a blank piece of white paper. What color image do you see now?

What Happened

When you looked at the white paper after staring at the blue dog, you saw the dog in yellow on the paper.

The retina of your eye includes two kinds of light-sensitive cells: rods and cones. The rods are very sensitive to light; your eyes use them to see at night, in dim light. The cones are the cells that are used for daytime vision, for the most part, and for seeing colors in bright light. If you stare at an object that is just one color, the cones are overstimulated, and when you look at the white paper you see all the colors except for the one you were staring at before. When the remaining colors combine, you get the complement (color opposite) of the color. Yellow is the complement of blue and it is the color you would see if red and green light combined (see the Primary Colors project).

GOING 'ROUND IN CIRCLES

The strangest thing happens when you put two regular grids or gratings on top of each other: you see a wavy image called a moiré pattern. Will your eyes go around in circles? Here are a few ways for you to try this experiment and see for yourself.

You Will Need

two hair combs

sheer curtain material

fine mesh window screen

What to Do

1. Place one comb in front of the other so that the two combs overlap and one is about ½ inch (1 cm) more distant from you than the other. Hold the combs about an arm's length away from your eyes. Look through the teeth of the combs, while moving the combs from side to side. Change the position of the combs so that they overlap in different directions. What happens to the patttern you see when the combs are moved?

2. Fold a piece of sheer nylon curtain or a similar material into two layers. Hold it about a foot (30 cm) away from your eyes and look through the material while moving one layer of cloth up and down. What happens?

3. Use two pieces of fine-mesh window screen and repeat the experiment done in Step 2. What happens?

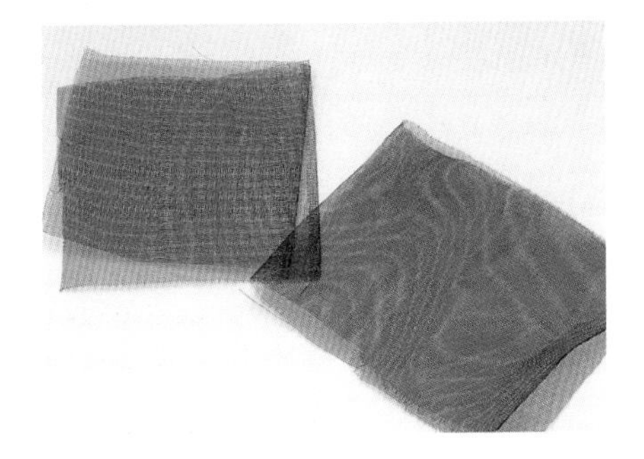

What Happened

Whenever two regular grids or gratings overlap one another, you get a moiré pattern—a wavy pattern of light and dark lines. Different patterns were created, depending on what the grids looked like, how fast you moved them back and forth, and how far away you were when viewing them. If the grids line up exactly, the moiré pattern doesn't occur. The pattern is an optical illusion caused by the light and dark regions of the screen or grating crossing each other. Small differences in the two designs cause the wavy pattern.

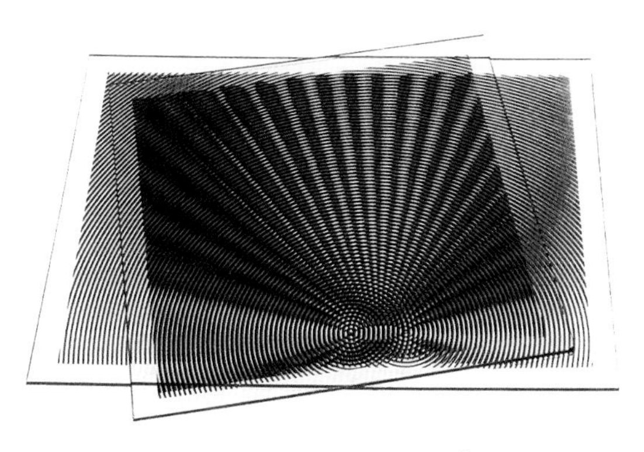

You can see a moiré pattern when you look through two overlapping grids.

Part Six

Optical Instruments

Instruments that use light can be called optical instruments. Optical instruments use lenses and mirrors (which actually are simple optical instruments themselves) to reflect and refract light. They let us see things that we couldn't see with our eyes alone.

You'd be amazed at how many everyday devices are based on optics. Here are some examples:

Camera: A camera is basically a light-tight box with a convex lens on the end.

Periscope: A periscope, which enables you to see things in a different direction than your eyes alone can see, is essentially two mirrors at an angle.

Telescope: Telescopes, first invented over 300 years ago, use lenses and mirrors to magnify things far away. Telescopes today are used on Earth and in space to send images back to Earth for astronomers to study.

Microscope: Microscopes let us see tiny things that we couldn't detect with just our eyes. Early microscopes could magnify objects about 250 times. Today, electron microscopes can magnify objects over 400,000 times!

Optical fibers: Hair-thin glass or plastic optical fibers allow people to look at things in a completely different way; they are changing the way we communicate.

Laser: A laser is an optical instrument that produces light of a single wavelength in which the waves are all traveling in step with each other. Lasers can be used to perform delicate surgery or cut holes in things with great precision. A laser beam reflecting off the bottom of a compact disk (CD) gives information that is translated into sound or images.

In this section of the book, we'll learn more about some simple optical devices, and make a few of our own.

Bundle of optical fibers.

LIQUID LIGHT

Skylights and windows are used to light the insides of houses. They give natural light and it doesn't cost anything. Wouldn't it be nice to be able to light all the rooms of your house with free natural light? Now you can. Scientists have developed a system to transport sunlight to any part of a building using optical fibers, hair-thin glass or plastic fibers. Here is an experiment to show you how these fibers work.

You Will Need

empty can (12 oz. size)

can opener

nail and hammer

black construction paper or aluminum foil

masking tape

water and bucket

flashlight

What to Do*

1. Remove the top of an empty can with a can opener. Discard the top. Put masking tape around the top edge of the can so it isn't sharp.

2. Hammer a nail through the side of the can near the bottom to make a small hole. Remove the nail.

3. Form a cone-shaped sleeve of black construction paper or aluminum foil around the front part of the flashlight (leave the switch clear). The open end of the cone should be wide enough to fit around the top of the can. Use the masking tape to hold the cone in place on the flashlight.

4. Tape up the nail hole in the can with a piece of masking tape. Fill the can with water.

5. Put the cone of paper over the can as shown in the photo. Turn on the flashlight and turn off the lights in the room. Hold the can over the bucket and remove the

**Ask an adult to help you with steps 1 and 2 of this project, if necessary. Remember that a cut can may have sharp edges.*

Make nail hole in can at bottom.

Flashlight with cone of paper in place over can.

masking tape from the side of the can. Make sure the stream of water goes into the bucket.

What Happened

You saw the light traveling in the stream of water into the bucket. Most of the light was

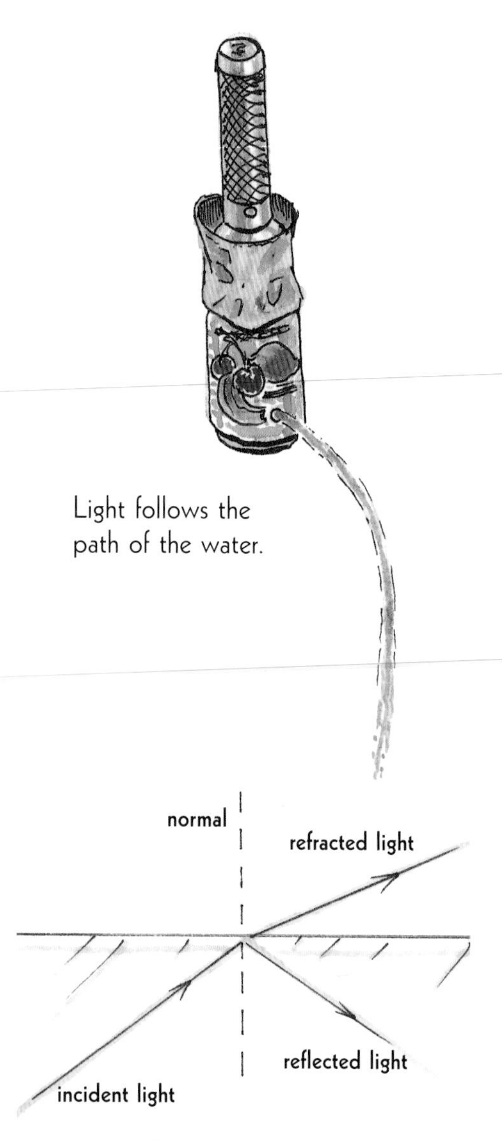

Light follows the path of the water.

1. Usually, part of light is reflected, part refracted.

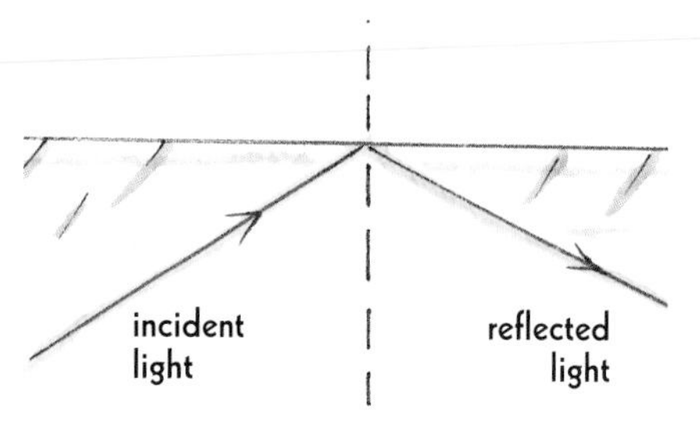

2. Total internal reflection.

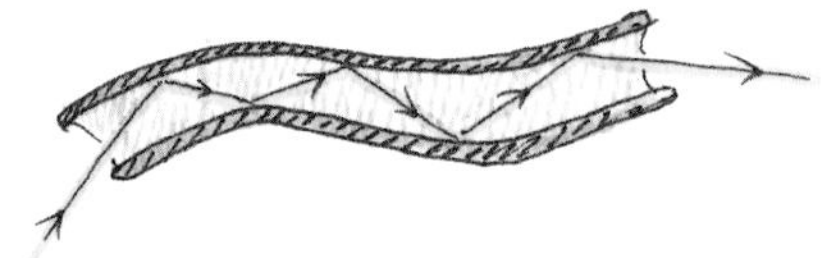

3. How light travels in an optical fiber.

reflected back and forth inside the stream of water. (Some escaped, which is why you could see it.)

When light strikes the boundary between two transparent materials, the light generally divides into 2 parts. Part of the light is reflected and part of the light is refracted (bent); see figure 1. We saw that this happened in water droplets earlier in our book when we studied rainbows. For the reflected light, the angle of incidence equals the angle of reflection, as we saw in our mirror study Against the Law.

When light passes from a medium with a larger index of refraction to a medium with a smaller index of refraction (for example, from water to air), the refracted ray bends away from the normal. It is possible to increase the angle of incidence to a certain angle, called the critical angle, so that all the light is reflected back into the medium with the larger incidence of refraction, and none passes out (see figure 2).

This is what scientists do with optical fibers (figure 3). Optical fibers have a hair-thin glass or plastic core with a high index of refraction and a covering with a lower index of refraction. Light is beamed into the fibers at such an angle that all of it bounces along inside

Light in special plastic rod with total internal reflection.

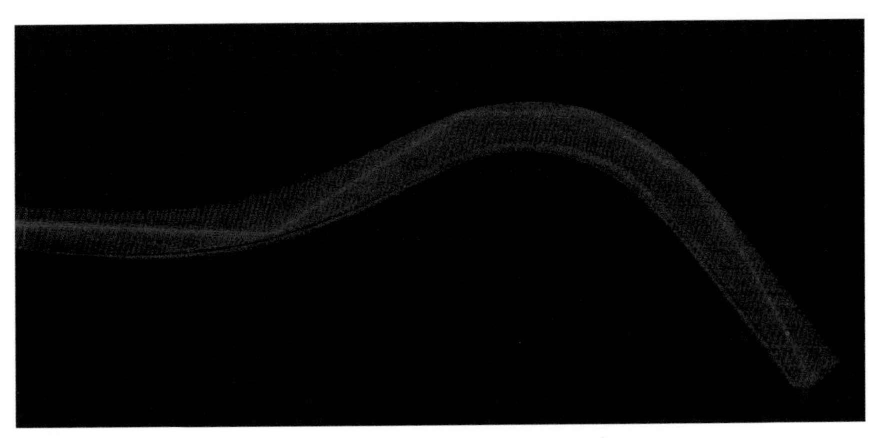

Light pathway in special plastic shows total internal reflection.

the fiber core and is reflected, without passing out. This is called total internal reflection. The light can be divided up to carry signals in communications networks. When you speak on the telephone, your voice is changed into an electrical impulse, which changes to a laser beam. This beam is fired along a fiber optic cable and then changed back into an electrical impulse, which is used to create the sounds your friends hear at the other end of the line.

Total internal reflection can be used to light rooms in buildings. The light travels down the fibers from the roof and is used to shine from specially developed light fixtures.

TWO-WAY MIRRORS

In many science museums, there are two-way mirrors. When the light shines on the mirror a certain way, you see only your reflection in the mirror. As you change the light, you begin to see what is on the other side of the mirror. Here is a way to examine how this works.

You Will Need

Mylar™ reflecting film from a potato chip bag or a Mylar balloon

damp cloth

light

What to Do

1. Cut a small square of Mylar about 4 inches × 4 inches (10 cm × 10 cm). Clean any food from the surface with a damp cloth.

2. Place the Mylar on a flat surface, with the shiny side upwards, and look at your reflection in it.

3. Hold the shiny side of the Mylar very close to your eyes and look through it at a bright light.

Hand can be seen through Mylar.

Cut a square of Mylar from a potato chip bag.

What Happened

When the Mylar was held close to your eyes and the light was very bright behind the Mylar, you could see the light through the Mylar. Mylar is a metal-coated flexible plastic sheeting. Metals reflect light and absorb very little light. This is why they look shiny. In the Mylar chip bag, the coating of metal is very thin. There is room between the tiny particles of metal on the plastic for light to pass through. This is the same thing that happens in two-way mirrors. They are thinly coated with metal. The side that is brightly lit acts like a mirror and reflects images back that are on the same side. The other side of the two-way mirror has lighting that is dimmer. If you are on the side of the mirror with the dim lighting, you can see through the mirror.

SMILE FOR THE CAMERA

A camera works a lot like an eye. The shutter is like the eyelid. Cameras and eyes both have lenses. And both focus light on a light-sensitive layer—the film in the camera and the retina in the eye. Usually the inside of a camera is black, to keep any light that enters the body of the camera from reflecting off the sides onto the film. Only the light that is focused on the film is supposed to affect the film. Here is a way to make a camera that uses a pinhole instead of a lens.

You Will Need

this book

scissors and craft knife

ruler

thick black marker

glue or glue stick

some lightweight cardboard, such as a cereal box

1-inch (2.5 cm) square of aluminum foil

straight pin

1-inch (2.5 cm) adhesive bandage (such as Band-Aid®)

masking tape

photographic paper (positive or negative)*

an adult

red or amber electric light bulb*

photographic developer*

photographic stop bath*

photographic fixer*

trays and equipment for developing prints*

darkroom or similar space that has no daylight coming in*

**These things are optional. If you can't get them, you can do the experiment using tracing paper instead of photographic paper. In this case you will see the image without being able to save a copy of it as a photographic image.*

What to Do

1. Photocopy the camera patterns so you have 2 each of piece A and piece B.

2. Cut out the photocopied

At back: Pattern for end of camera. Left: The assembled end. Right: The open box, assembled.

pieces of the camera patterns along the outer, solid lines. If the pieces are not completely black, use the marker to darken them.

3. Glue the two A pieces together so that they form a 6 1/8 × 11¾ inch (15.5 × 30 cm) rectangle, by gluing the narrow flap of one A piece behind the end of the second A piece (see figure 1, page 66).

4. Glue the rectangle you made in Step 3 onto a piece of lightweight cardboard with the black side facing out. Cut out the rectangle from the cardboard. With a craft knife, lightly score the side of the cardboard with the paper along the dashed lines so it will bend easily. Use a ruler to guide your knife. Ask an adult to help you if necessary.

5. Fold the rectangle along the dashed lines to make an open box shape (black side in) and glue the end flap of the rectangle to the opposite side of the box as shown in figure 2. This will be the body of the camera.

6. Glue a photocopy of piece B onto a piece of lightweight cardboard and cut it out along the outer lines. Score along the dashed lines with a craft knife as you did in Step 4. Cut up to the center rectangle on the dotted lines only (see figure 3). Fold the flaps and glue them in place (as shown on figure 4) to form a camera end. Make another camera end the same way.

7. With a craft knife, cut out the small dashed rectangle from *one* end piece of the camera and tape a piece of aluminum foil over it on the black side. Pierce the foil once carefully with a pin to make the pinhole. Place the plastic bandage lightly over the hole on the outside (cardboard side) of the end to act as a shutter. Glue or tape this camera end onto one end of the camera body.

8. Get an adult to help you with the rest of this project. Turn on a red or amber light in a completely dark room. Be

1. Glue one copy of Piece A to the second Piece A so the flap overlaps.

2. Glue the flap on the joined A pieces to make an open box.

3. Cut Piece B along the dotted lines to form flaps on the ends.

cut

cut

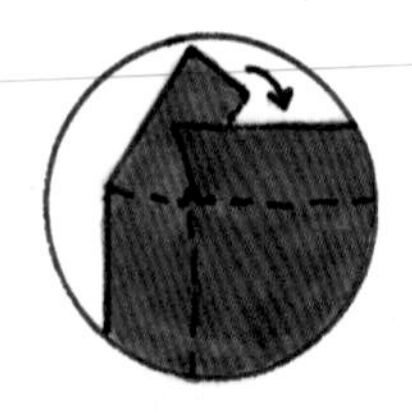

4. Glue the flaps on the ends.

5. The finished end.

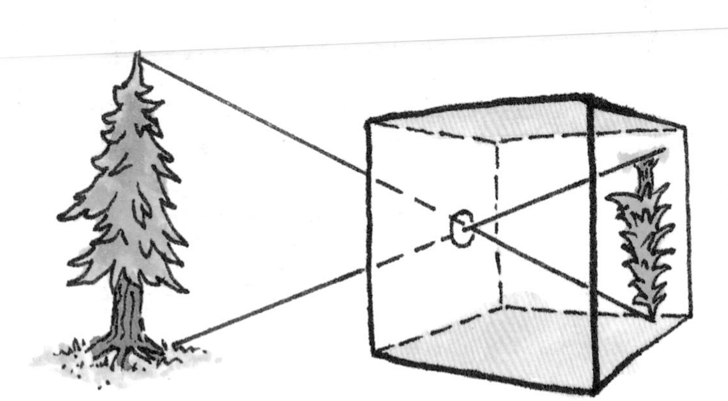

6. How an image is formed.

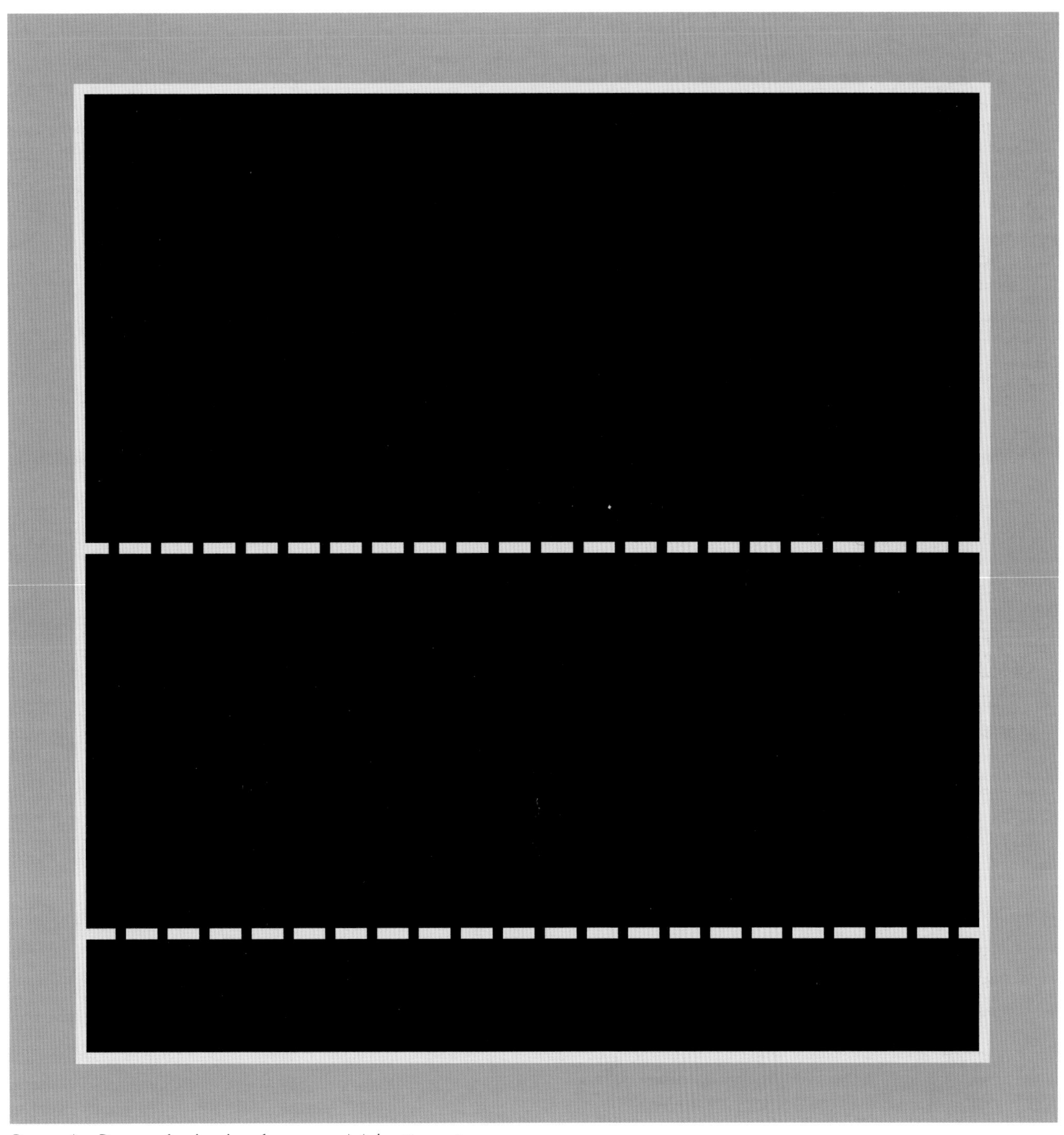

Piece A: Pattern for body of camera. Make 2 copies.

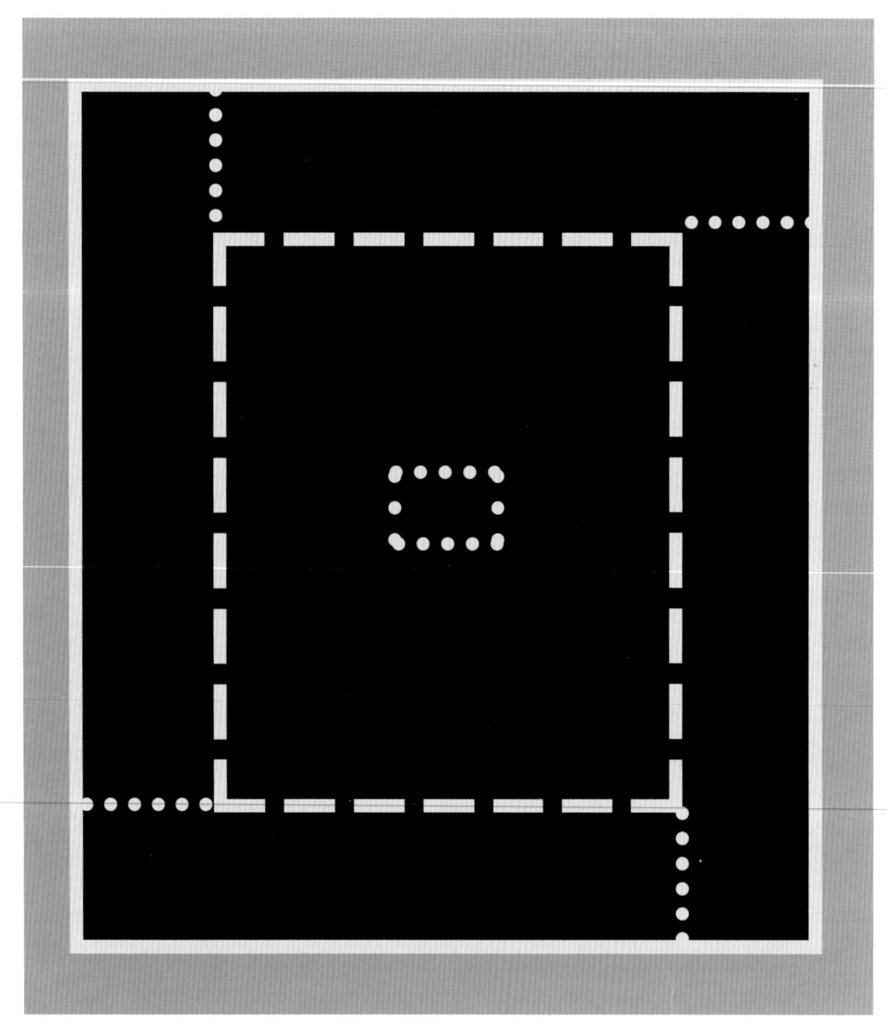

Piece B: Pattern for end of camera. Make 2 copies.

sure no light is coming into the room from outside. Remove a piece of photographic paper from its lightproof envelope and cut a 2¼ × 2¼ inch (6 × 6 cm) piece to fit the end of your camera. Carefully tape the piece of photographic paper onto the inside of the separate end piece of the camera, with the paper's emulsion side (shiny side) facing out. You may have to practice this with a discarded piece of photographic paper first. Tape the end of the camera in place on the camera body. Be sure all the seams are sealed with tape. Once the box is light-tight, it is safe to turn on the regular lights.

9. Take the camera to the location where you wish to take a picture. Carefully remove the plastic bandage from the pinhole. Keep the camera very still. The light from the pinhole will fall on the photographic paper, exposing the paper. After two or three minutes, replace the plastic bandage.

10. You can return to the darkroom and place the paper in a lightproof envelope like the envelope the paper came in, and repeat steps 8 and 9 several times to take several pictures. The amount of time required for an image to form on the photographic paper will depend on the size of the pinhole. Try different times from 30 seconds to 5 minutes, to see which works best with your camera.

11. When you have taken a few pictures, return to the darkroom. Have an adult help you mix the developer, stop bath, and fixer according to the packages' directions, and follow the development procedure suggested for your paper. Pay careful attention to the safety precautions and disposal information which comes with these chemicals.

Note: If you cannot get photographic chemicals or use a darkroom, you can cut a 2 × 2

Pinhole camera and some negatives made with it.

inch (5 × 5 cm) square out of the unattached camera end piece and tape a slightly larger piece of tracing paper over the hole. Hold the pinhole up to a bright object. The image of the object should appear on the tracing paper.

What Happened

You made a pinhole camera. The light rays enter the pinhole and are focused onto the sheet of photographic paper at the back of the camera. The image of the object you photographed was upside down inside the camera. Figure 6 (p.66) shows how the image is formed by the light rays entering the camera. The best images are usually made when the object you are photographing is not moving and is a few feet away. If you used normal photographic paper, the image will be a negative. The dark areas, such as shadows, will appear light and the light areas (such as light bulbs or reflections from glass) will appear dark. If you used direct positive paper, you will have a positive image; the objects that look light to your eyes will appear light in your print.

CLOSER AND CLOSER

There are tiny fossils in your toothpaste, which help remove the stuff growing on your teeth. The yogurt you eat is filled with living bacteria. (Don't worry; they won't hurt you; they helped make the yogurt.) There may be tiny dust mites under your bed. Why can't you see them? They're too small. But you can make them look bigger. Welcome to the world of microscopes! Microscopes are used in medical labs, crime labs, research, and manufacturing. People look at blood cells, microchips, and diamonds with microscopes.
Here's a way to make a very simple microscope.

You Will Need

stiff cardboard

small needle

scissors

water

newspaper or other paper with writing on it

lamp

white corn syrup (optional)

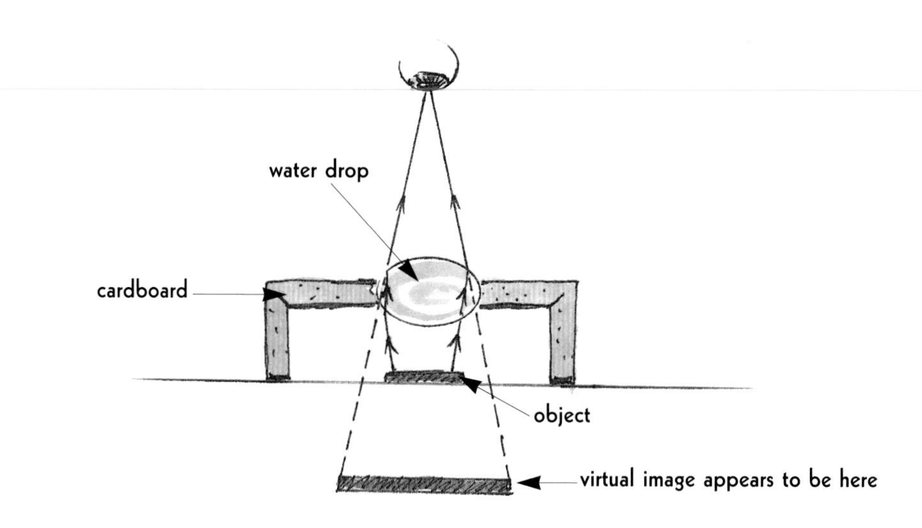

Diagram of microscope, seen from the side. The size of the water drop is exaggerated to show the light rays.

What to Do

1. Cut a piece of cardboard 6 inches × 1½ inches wide (15 cm × 4 cm). Draw a line 1 inch (2.5 cm) in from each short end and fold the ends to form "legs" for the microscope.

2. Have an adult use a needle to poke a tiny round hole in the center of the cardboard, about an equal distance from each end. This serves as the eyepiece of the microscope.

3. Place a tiny drop of water in the eyepiece. If you are careful, it will not drip through the hole.

4. Place your microscope on top of a newspaper. Bring a lamp a few inches away from your newspaper, so the light shines onto it and it is brightly lit. Look through the eyepiece to view the letters. Raise and lower the eyepiece by bending the legs of the microscope more or less. How does this affect the clarity of the letters on the newspaper?

5. Try making another pinhole in the microscope. This time use white corn syrup instead of water for the lens.

What Happened

You made a very simple microscope. Both the drop of water and the drop of corn syrup were thicker in the middle than on the edge, so each acted like a convex lens and focused the images of the letters that were on the newspaper. The image you saw through the drop was a virtual image. It was right-side up and appeared to be larger than the original print. The corn syup may have enlarged the letters more than the water drop.

Corn syrup is more viscous and forms a fatter lens than water; it also bends light more than the water does.

The simplest kind of microscope is a simple convex (converging) lens—what you may know as a magnifying glass. Modern microscopes are usually compound microscopes. They have more than one lens. They can magnify the objects so that they appear to be hundreds or even thousands of times larger than life size.

TERRESTRIALS

If E.T. was an extraterrestrial, a creature who wasn't from Earth, there must be terrestrials. That's us! All the things that live on Earth are terrestrials. Perhaps in a galaxy far, far away, there are aliens looking at us. Telescopes help us look at distant objects on Earth or in space. In this project, we'll make a simple telescope.

You Will Need

2 cardboard tubes, one slightly wider than the other*

thin convex lens about 1½ inches (4 cm) in diameter

thick convex lens about 1½ inches (4 cm) in diameter

piece of thick corrugated cardboard about 6 × 6 inches (15 × 15 cm)

pencil

drawing compass

scissors

craft knife

yarn

masking tape

**An empty paper towel tube and an empty plastic wrap tube work well. The diameter of the thinner tube should be the same as the diameter of the thin lens. The diameter of the wider tube should be slightly larger than the diameter of the thick lens. If you don't have tubes, use two pieces of thick paper or thin cardboard about 8½ × 11 inches (21.5 × 28 cm).*

1. Take a cardboard tube with an opening about the same diameter as the thin convex lens. If you don't have a tube the same diameter as the lens, you can make your own by rolling a thick piece of paper or thin cardboard into a tube of the correct width and then taping it around to keep it rolled. The finished tube should be about 11 inches (28 cm) long. Tape the thin lens in place on its edges over one of the ends of the tube. Set it aside.

2. Trace a circle the same size as the opening of the wider tube, or slightly wider, onto the corrugated cardboard and have an adult cut it out. This will be the holder for the thick lens, which will be the eyepiece of the telescope.

3. Trace a circle in the center of the holder that is about 1 inch (2.5 cm) smaller than the diameter of the holder. The center circle should have a smaller diameter than the lens. Cut the holder in half, so that the circles you drew are evenly divided (Figure 1).

4. Have an adult help you cut away the two halves of the center circle with a craft knife (Figure 2). If you put the two halves back together again, the holder would look like a disk with a hole in the middle.

5. Inside each half of the circle, use your scissors point to carefully break the cardboard corrugations between the top and bottom cardboard layers that surround the hole so the thick lens will fit snugly between the half-circles (Figure 3). Push the thick lens into each side of the holder and join the two halves of the holder with tape (Figure 4). You now have a holder with a thick lens in the middle (Figure 5).

6. Tape the holder with the thick lens to one end of the wider tube.

7. Wrap enough yarn around the middle of the narrower tube so that the tube can slide easily within the large tube, but won't slip out (Figure 6). Tape the yarn in place.

8. Insert the lens end of the narrower tube in the open end of the wider tube (Figure 7). Point the open end of the nar-

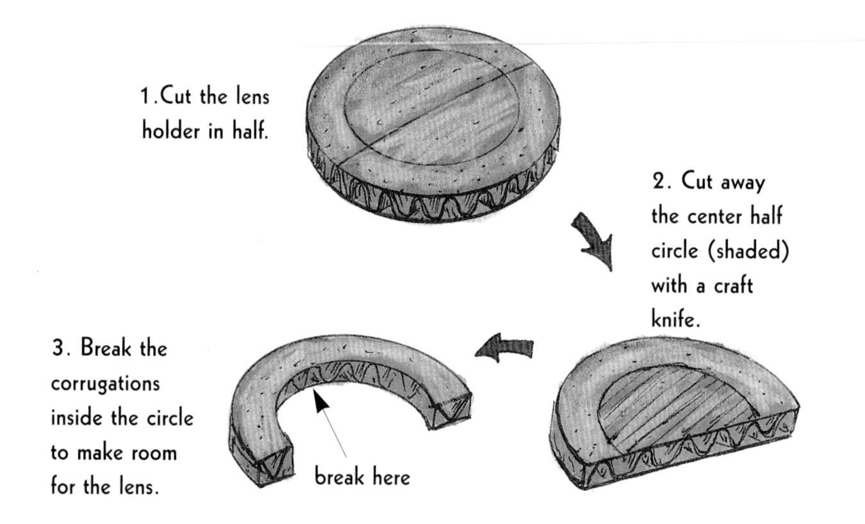

4. Push the thick lens into each side of the holder.

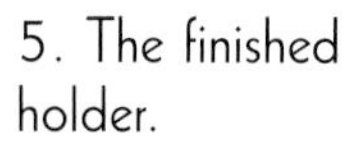

5. The finished holder.

increase the distance between the two lenses.

What Happened

You created a refracting telescope. It uses lenses to focus and collect light. The thinner lens functions as an objective. It focuses a real image of a faraway object; the image is small and upside down. The thicker eyepiece lens magnifies the image so that you can see it. Moving the tube back and forth allows you to determine the best distance apart for the lenses so that the image is in sharp focus. **Reminder: Do not point your telescope at the Sun.**

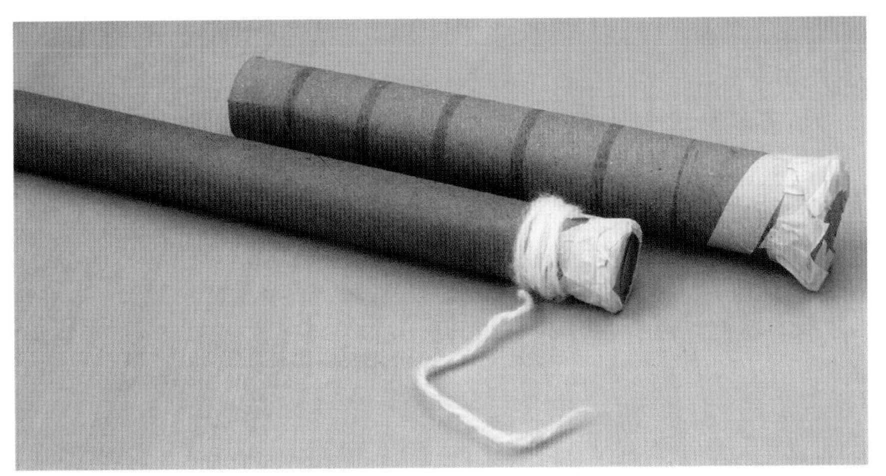

6. Finished tubes, with yarn wrapped around the narrow tube.

7. Assembled telescope.

row tube at an object and look at it through the lens of the outer tube. Move the inner tube back and forth to get the best image. If you can't get a clear image, try reversing the position of the inner tube so the lens end is facing out, to

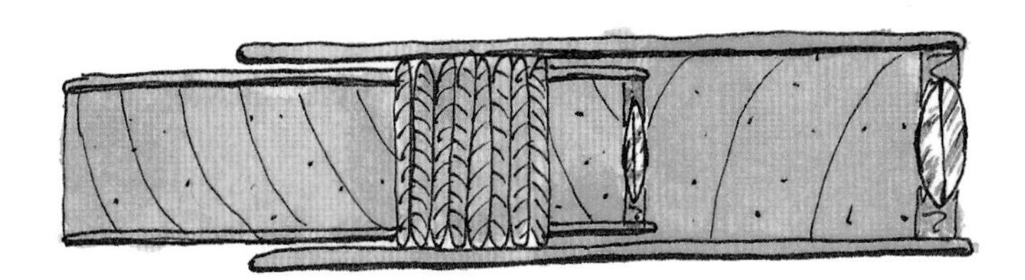

8. Two possible ways of assembling the telescope.

MOONBEAMS

People throughout time have been fascinated by the Moon. Pictures of the Moon have appeared in ancient paintings, and every culture in the world has some legend or practice associated with the phases of the Moon. Here is an experiment that will allow you to study the Moon. Reminder: Do not point your telescope at the Sun.

You Will Need

a clear night with a Moon

window

a magnifying mirror (shaving or makeup mirror)

large, flat mirror

tape (optional)

large magnifying glass

What to Do

1. Put a table near a window, where you can get a good view of the Moon. Turn off the lights in the room. Angle the magnifying mirror so that it faces away from you and towards the image of the Moon.

2. Place the large flat mirror against the sill of the window, and adjust the mirror until the light from the Moon reflected by the magnifying mirror can

be seen in the flat mirror. It will take some time to get the two mirrors at the correct angle.

3. When you can see the moonlight in the flat mirror, hold a magnifying glass at arm's length away from you and towards the Moon's image in the flat mirror. Bring the magnifying glass towards you until you can view the Moon through the magnifying glass. Is the image that you see upside down or right-side up?

What Happened

You have created a reflecting telescope. The concave mirror produced a real image (upside down) of the Moon, which was reflected by the flat mirror. The concave lens of your magnifying glass focused the real image of the Moon so you could see it clearly.

The first successful reflecting telescope was made in 1668 by Sir Isaac Newton. Today reflecting telescopes are used by astronomers to look at objects in space. The Hubble Space Telescope is a reflecting telescope.

SPY HARD

Here is a fun way of exploring optics while peeking around corners at the same time. A periscope is an instrument used on a submarine to help the captain see what is going on above the water without having to bring the sub to the surface. You won't have to go under water to use this periscope: hide under a blanket and see who is sneaking into your room!

You Will Need

one or two empty quart- or litre-sized cardboard milk or juice containers

2 small rectangular mirrors

tape

craft knife

What to Do

1. Wash and dry the container thoroughly.

2. Have an adult cut a rectangular cutout slightly smaller than the mirror near the bottom of one side of the container (see diagram, p.76).

3. Have an adult cut another cutout, about the same size as the first one, on the opposite side of the container from the first cutout, near the top of the container.

4. Place a mirror inside the container through the bottom cutout and tape it to the container with the mirror side up at a 45-degree angle to the bottom, so that it points upward towards the top of the contain-

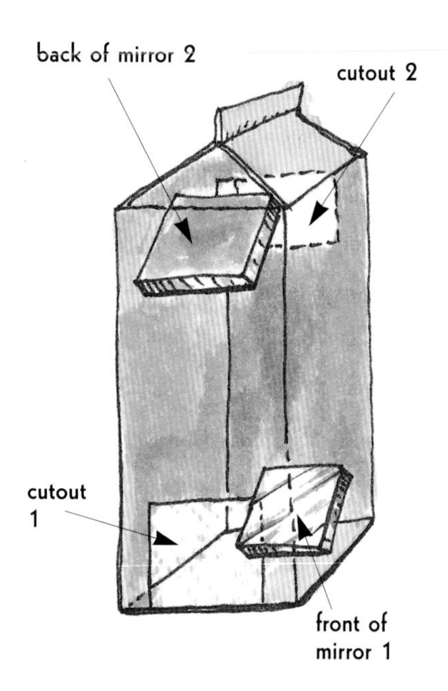

In this diagram we pretend we can see through the walls of the periscope.

er (see diagram). (Note: If you want to, carefully cut open another long side of the container to aid in placement of the mirrors, as shown in the photo, and then tape it closed when you are finished.)

5. Place the second mirror near the top of the container at a 45-degree angle to the top, so that it is parallel to the other mirror. The mirror side of the top mirror should be facing down. Tape the mirror to the sides of the container.

6. Point the top cutout in the direction that you want to look, and peer through the bottom cutout. What do you see?

Cut the side of the container open if you want to, to make it easy to attach the mirrors. Retape it closed later.

7. If you want a longer periscope, add a second carton. Cut the top and bottom off one carton, attach it to the first carton, and make an opening as you did in Step 5.

What Happened

You have created a periscope. With a periscope you can see over or around an obstacle. If you can raise the top of your periscope over a fence, you can see over it, even if you're not tall enough to look over it yourself. Periscopes in submarines have gone high-tech and don't necessarily look like the ones you see in old movies. They now have t.v. cameras or optic fibers that show what is happening on the surface. There are also special zoom lenses which will enlarge or focus in on a certain area. There are markings on the viewer to tell how far away an object is, what direction it is traveling in, or even how large the object appears.

GLOSSARY

absorption: changing of electromagnetic energy to other forms of energy as it passes through a medium

additive primary colors: red, blue, and green light, which combine to give all the other colors and white when mixed in the correct amounts

angle of incidence: the angle an incident beam of light makes with a surface, measured from the normal (a line perpendicular to the surface)

angle of reflection: the angle a reflected beam of light makes, measured from the normal, as it leaves a surface

angle of refraction: the angle a refracted beam of light makes, measured from the normal, as it travels from one medium to another

baffle: a device for stopping the passage of, turning, or regulating the amount of light

birefringence (or double refraction): property of certain materials of forming two refracted light rays, which are differently polarized, from a single ray

blind spot: the area on the retina where the optic nerve enters the retina. It has no receptors and so it is insensitive to light

color blindness: a condition in which a person is unable to detect or distinguish between certain colors, usually red and green, because he or she is missing certain color receptors in the retinas of the eyes

concave lens: a lens that is thicker at the edges and thinner in the middle. Also called a diverging lens.

concave mirror: a mirror curved like the inside of a bowl

converging lens: a lens that is thinner at the edges and thicker in the middle. Also called a convex lens.

convex lens: a lens that is thinner at the edges and thicker in the middle; a converging lens.

convex mirror: a mirror curved like the outside of a bowl

critical angle: the angle at which all light is reflected within a transparent substance; no refracted light emerges

diffraction of light: the bending or spreading out of a light beam as it moves around an object or through a slit. Diffraction produces fringes of parallel light-and-dark or colored bands

diffraction grating: a piece of glass or plastic with many close parallel thin lines used for breaking light up into spectra (plural of spectrum) by diffraction

diffuse reflection: scattering of a beam of light by a rough surface

electromagnetic spectrum: the complete range of electromagnetic waves from the longest wavelength (radio waves) to the shortest wavelength (gamma rays)

fringes: Pattern of light-and-dark stripes or colored stripes made by a diffraction grating or other material that diffracts light

hydroelectricity: electricity produced using water power

image: a representation of an object formed by a lens, mirror, or other optical instrument.

incident: (adj). falling on or striking

index of refraction: the ratio of the speed of light in a vacuum to its speed in a particular substance

lens: a curved piece of polished glass, plastic, or other transparent material used for refraction (bending) of light

light: the portion of the electromagnetic spectrum visible to the eye

mirage: an optical occurrence caused by refraction of light through layers of air that have very large temperature differences

mirror: a surface that reflects most of the light that falls on it

mirror image: the view of something as if seen in a mirror

models: devices or ideas used to help understand or explain an occurrence

nanometer (nm): 1/1000,000,000 of a meter. Wavelengths of light are measured in nanometers.

normal, the: a line at right angles to (perpendicular to) a surface

optical density: the property of a medium that determines the speed of light in that medium

optics: science that studies light, vision, and optical instruments

plane: a flat surface on which any two points can be joined by a straight line

polarization: a process of affecting light so that the vibrations have a definite direction, such as along the same plane or spiral

prism: a block of transparent material such as glass or plastic, usually having triangular bases and sides that are parallelograms; a prism is used for refracting light

ray: a narrow beam of light

ray box: a device that produces parallel light beams

ray model: a model of light that represents light beams as rays moving in straight lines outwards from a source

real image: an image is a representation of an object formed by a lens, mirror or other optical instrument. If the rays of light coming from the image actually pass through the image, it is called a real image. Real images are inverted and can be projected onto a screen

reflection: the return of all or part of a beam of particles or waves when it meets the boundary between two substances

refraction: the bending of a light beam that occurs at the boundary between one medium and another, when light strikes it at an angle other than 90 degrees

spectrum: range of electromagnetic energies in order of increasing or decreasing wavelength. The visible spectrum of light ranges from red light (700 nm) to violet light (400 nm)

specular reflection: the reflection of light in a beam, as from a mirror

stress birefringence: birefringence which occurs when some types of material are stressed or compressed (see birefringence)

subtractive primary colors: primary colors of paint which combine to produce all the colors, including black. Typical subtractive primaries are cyan, magenta, and yellow.

total internal reflection: the total reflection of a beam of light at the boundary of a medium (such as glass) and another medium with a lower index of reflection (such as air); no light emerges

transmitted light: light that passes through transparent materials

virtual image: if an image is seen at a place from which light rays appear to come to the viewer, but the light rays actually aren't coming from that place, the image is called a virtual image. Virtual images cannot be projected onto a screen. The image you see in a flat mirror, for example, is a virtual image. The image appears to be coming from behind the mirror.

visible spectrum: the section of the electromagnetic spectrum that can be detected by the human eye

vision: the act of seeing

wavelength: the length of a wave, measured between the same points on two waves that are next to each other; for example, from the highest point (crest) on one wave to the crest on the next wave

INDEX